65th Conference on Glass Problems

Editorial and Circulation Offices
PO Box 6136
Westerville, Ohio 43086-6136

Contact Information
Customer Service: 614-794-5890
Fax: 614-794-5892
E-mail: info@ceramics.org
Website: www.ceramics.org

Ceramic Engineering & Science Proceedings (CESP) (ISSN 0196-6219) is published nine times a year by The American Ceramic Society, PO Box 6136, Westerville, Ohio 43086-6136; www.ceramics.org. Periodicals postage paid at Westerville, Ohio and additional mailing offices.

Change of Address: Please send address changes to *Ceramic Engineering and Science Proceedings,* PO Box 6136, Westerville, Ohio 43086-6136, or by e-mail to info@ceramics.org.

Subscription rates: One year $295 (ACerS member $236) in North America. Add $40 for subscriptions outside North America. In Canada, add GST (registration number R123994618)

Single Issues: Single issues may be purchased online at www.ceramics.org or by calling Customer Service at 614-794-5890.

Indexing: An index of each issue appears at www.ceramics.org/ctindex.asp.

Contributors: Each issue contains a collection of technical papers in a general area of interest. These papers are of practical value for the ceramic industries and the general public. The issues are based on the proceedings of a conference. Both The American Ceramic Society and the non-Society conferences provide these technical papers. Each issue is organized by an editor(s), who selects and edits material from the conference proceedings. The opinions expressed are entirely those of the presenters. There is no other review prior to publication. Author guidelines are available on request.

Postmaster: Please send address changes to Ceramic Engineering and Science Proceedings, PO Box 613, Westerville, Ohio 43086-613. Form 3579 requested.

65th Conference on Glass Problems

A Collection of Papers Presented at the 65th Conference on Glass Problems, The Ohio State University, Columbus, Ohio (October 19-20, 2004)

Editor
Charles H. Drummond, III

Published by

The American Ceramic Society
PO Box 6136
Westerville, Ohio 43086-6136
www.ceramics.org

65th Conference on Glass Problems

For information on ordering titles published by The American Ceramic Society, or to request a publications catalog, please call 614-794-5890, or visit www.ceramics.org

ISSN 0196-6219

ISBN 1-57898-238-9

Cover image: Glass art by Dale Chihuly from his recent exihibition at Franklin Park Conservatory in Columbus, Ohio. Photographer Terry Rishel.

Contents

Foreward

The 65th Conference on Glass Problems was sponsored by the Departments of Materials Science and Engineering at The Ohio State University and the University of Illinois at Urbana-Chapaign. The director of the conference was Charles H. Drummond, III, Associate Professor, Department of Materials Engineering, The Ohio State University. Dean William A. Baeslack, Colllege of Engineering, The Ohio State University, gave the welcoming address. Chairman John Morral, Department of Materials Engineering, The Ohio State University, gave the Departmental welcome.

The themes and chairs of the four half-day sessions were as follows:

Processing

Ruud G. C. Beerkens, TNO-TPD, Eindhoven, The Netherlands
Robert Thomas, Corning, Corning, NY

Conditioning, Refractories and Combustion

Daryl S. Clendenen, Vesuvius-Monofax, Falconer, NY
Tom Dankert, Owens-Illinois, Toledo, OH

Furnaces

C. Philip Ross, Glass Industry Consulting, Laguna Niguel, CA
Robert Thomas, Corning, Corning, NY

Future of the Glass Industry

Robert Lawhon, PPG Industries, Pittsburgh, PA
Larry McCloskey, Toledo Engineering, Toledo, OH

An off-the-record panel discussion on the future of the glass industry followed the presentations.

Presiding at the banquet was Professor Waltraud M. Kriven, Department of Materials Science and Engineering, the University of Illinois at Urbana-Champaign. The banquet speaker was Jay A. Kandamppully, Professor of Consumer and Textile Sciences, The Ohio State University. His address was titled "Competing in the Product Market Through Services."

The conference was held at the Fawcett Center for Tomorrow, The Ohio State University, Columbus, OH USA.

Preface

In the tradition of previous conferences, stated in 1934 at the University of Illinois, the papers presented at the 65th Annual Conference on Glass Problems have been collected and published as the 2004 edition of The Collected Papers.

The manuscripts are reproduced as furnished by the authors, but were reviewed prior to presentation by the respective session chairs. Their assistance is greatly appreciated. C.H. Drummond did minor editing with further editing by The American Ceramic Society. The Ohio State University is not responsible for the statements and opinions expressed in this publication.

Charles H. Drummond, III
Columbus, OH
December 2004

Acknowledgements

It is a pleasure to acknowledge the assistance and advice provided by the members of the Program Advisory Committee in reviewing the presentations and the planning of the program:

Ruud G. C. Beerkens—TNO-TPD

Daryl S. Clendenen—Vesuvius-Monofrax

Dick Bennett—Johns Manville

Tom Dankert—Owens-Illinois

Robert Thomas—Corning

Robert Lawhon—PPG Industries

Larry McCloskey—Toledo Engineering

C. Philip Ross—Glass Industry Consulting

Advanced Feeder Control Using Fast Simulation Models

Oscar Verheijen, Olaf Op den Camp and Ruud Beerkens,
TNO Glass Group, The Netherlands
Ton Backx, IPCOS Netherlands, The Netherlands
Leo Huisman, Eindhoven University of Technology, The Netherlands

Abstract

For the automatic control of glass quality in glass production, the relation between process variable and product or glass quality and process conditions/process input parameters must be known in detail. So far, detailed 3-D glass melting simulation models were used to predict the effect of process input variables, such as fuel consumption and fuel distribution, load and load distribution, and electrical boosting, on the flow pattern (residence times, short cut flows), temperature distribution and redox state of the glass in the furnace. For feeder control, the main objectives are stable temperature and temperature uniformity in the spout section of the feeder just before the glass melt is delivered to the forming process. However, computations of detailed 3-D simulation models are very time-consuming: one steady state simulation of a complete furnace including refiner(s) typically takes about a day. This time demand indicates that the currently used detailed 3-D simulation models are not suitable for rigorous (CFD) model based Model Predictive Control (MPC). To make the 3-D simulation models suitable for control purposes, simulation tools that are much faster than real time with still a high level of reliability are now developed. These fast simulation tools (GPS: Glass Process Simulators) open up a wide variety of applications of glass furnace models: process monitoring, what-if scenarios and model based predictive control (MPC) of temperatures or glass melt quality or redox/color. Now we realized time-transient simulations with GPS for feeders, which are about 10000 times as fast as real time. This paper will show the capability of an MPC that is based on fast GPS to control temperature and temperature uniformity of one of the feeders connected to an emerald-green container glass furnace.

Introduction

At the moment, most feeders or forehearths of glass-melting furnaces are controlled manually. Feeder control is usually based on manual adjusting temperature set-points in different zones of the feeder. Mostly, these zonal temperature set-points are set such that a weighted average of the temperature readings of the 9-grid thermocouple arrangement close to the spout entrance reaches the desired value and to obtain minimum differences in the 9 couple readings. By adjusting the zonal temperature set-points, the operator adapts the total amount of fuel (or more specifically the total amount of energy in case also energy is supplied by means of electrical boosting) that is being used to realize and maintain the desired glass melt temperature (uniformity) at the feeder exit. As the temperature distribution in the glass melt throughout the feeder depends on feeder pull rate and glass melt redox state, the desired glass melt temperature in the feeder strongly depends on the product type (product type weight, color, associated pull) that is produced. Therefore, during product changeover, it is the task of the operator to modify the feeder zone temperatures such that the new desired glass melt temperature is reached within the shortest period of time and stable production can be continued. In general, feeder control strategy is based on operators' experience and is often operator dependent. For reproducible feeder control, automatic control of glass melt temperatures (or any other variable that determines feeder performance) is almost indispensable and can lead to considerable energy savings.

In the recent years, process control systems have become available for automatically control of important temperatures, taking over this part of the job of an operator, who can spend his/her time now on other important tasks such as furnace maintenance. Since PID controllers are much to slow and have no predictive capability (i.e. do not give feed forward response on disturbances to keep the temperature within a limited range), model based predictive control (MPC) is being used. MPC is dedicated software running on a regular PC that computes the required control actions (to obtain desired temperatures or other process variables) and communicates these control actions to the DCS (distributed control system) of the furnace by changing the set-points of the conventional controllers that are available within the DCS. For glass furnace feeders, the models in MPC describe the temperature changes in the feeder when changes in process settings such as pull, fuel input over the zones, and/or electrical boosting are made. A reverse response model is used to determine how to adapt the individual inputs (process settings) of the feeder to ensure temperature control as close as possible to the desired levels even for changeovers (following a pre-defined temperature course). Although conventional MPC systems become more and more accepted in the glass industry, their advantages with respect to manual control is limited, as they do not change the basic control strategy of keeping, in this case, the weighted average of the temperature readings of a 9-grid thermocouple arrangement at the feeder exit within a limited range. Derivation of so-called black box models for these conventional MPC controllers requires industrial tests. By interference in the actual glass-melting tank and/or feeder, flow pattern and temperature distributions are disturbed, which sometimes offsets the process and may lead to increased project reject and expensive man-hours to re-stabilize the process. Besides this, due to the increased computational power of PC's, MPC allows for more sophisticated control in which not one single (averaged) temperature is kept stable, but a total crown profile or melter bottom temperature profile is prescribed (MIMO = multi input – multi output).

For MPC, the simulation models that are used have to be extremely fast (over 1000 times real time) compared to the real process. Conventional MPC systems for feeders therefore make use of experimentally derived models that are determined by means of step tests on the feeder in which the inputs (i.e. fuel distribution, boosting, and load) are changed stepwise. The responses of the feeder (readings from e.g. thermocouples) are measured and correlated to the changes in input.

Herewith, models are derived that directly couple the (expected) response of the feeder to changes in process settings. Usually, the stepwise changes in process settings are chosen very small, as to prevent any disturbance of the feeder. Consequently, the linear models that are determined in this way have a limited area of application. Actually, the model can only be used when process settings do not deviate much from the settings at which the model is determined (the working envelope of the model). When this model is used in a situation outside its working envelope, the controller performance will decrease. In that case, the controller will either determine control actions that are less accurate and slow down compensation of a deviation between the controlled value and its set-point or it might even lead to contra-productive control actions or oscillation which leads to severe system instability. MPC controllers should be designed in such a way that system instability does not occur under any circumstances.

To overcome the limited working envelope of conventional MPC controllers, the use of detailed 3-dimensional computational fluid dynamics (CFD) models (white or rigorous models) is required. About 20 years after the introduction of detailed modeling of flows and temperatures in glass melting furnaces, simulation tools are now used for process design and optimization on a day-to-day basis in several sectors of the glass industry (special glass, glass fibers, TV glass, float glass, container glass)[1]. First of all, simulations are used to obtain a basic understanding of the interaction between the different processes that simultaneously take place in a glass-melting furnace. To determine the impact of changes in design or changes in process settings on the process (or more specifically on the expected glass melt quality), the simulation tools determine indicators to judge the performance of a furnace with respect to melting, fining, refining, and homogenization. Currently, most furnaces are only (re-) designed after evaluation of the expected furnace performance by means of simulation studies. More and more, simulation studies are used to determine possible solution strategies in case of sudden furnace upsets. In principle, CFD models have no limitations with respect to process settings. Therefore, these models are able to cope with the prediction of the dynamic behavior of a glass-melting furnace and feeder over a large working area (large variations in pull rate, color-changes caused by wide changes in glass melt redox state). Such a wide working range and off-line development of the control models makes these CFD models attractive for predictive control purposes in glass melting processes. However, computations of detailed 3-D simulation models are very time-consuming: one steady state simulation of a complete furnace including refiner(s) typically takes about a day, which indicates that the currently used detailed 3-D simulation models are not suitable for MPC. This was one of the reasons to develop simulation tools that are much faster with a high level of reliability. These fast simulation tools (GPS: Glass Process Simulators) open up a wide variety of applications of glass furnace models.

In the framework of a Dutch EET-project, TNO Glass Group and IPCOS in cooperation with the Eindhoven University of Technology have developed a method to set-up fast Glass Process Simulators and based on these simulators, MPC-controllers that enable control for a very large working envelope. In a first step, a fast process simulator (GPS) is derived from a validated 3-D detailed CFD model for the process of consideration (glass feeder or furnace). The resulting GPS is applied to an emerald green container glass furnace of REXAM in Dongen, The Netherlands ([2]: here GPS is used to monitor on-line all relevant process variables and to predict future process performance when changing input parameters). In a second step, an MPC controller was built on the basis of GPS and successfully applied to one of the three feeders of the above-mentioned furnace [3]. The feeder was selected because of the large job changes that are regularly applied to it.

MPC Control Based Upon Detailed CFD Models

The following approach to derive fast simulation models for MPC control was proposed and put into practice to set-up an MPC controller for an industrial feeder. The objective of the controller is to stabilize the temperature of the glass melt that is delivered to the forming machines in order to improve the so-called workability of the glass and to enable a controlled forming processes and optimum glass distribution in the mould. The workability of glass, or the ease with which the glass can be used for forming the final product, depends largely on the viscosity and therefore on the temperature and temperature distribution of the glass melt.

A new approach to set-up control models:

- The approach starts with setting up a separate GPS based on a detailed CFD model for the feeder under consideration. This GPS is of course thoroughly evaluated and validated, as the performance of the controller will depend largely on the quality of the original model.
- Subsequently, dynamic tests are performed upon the GPS model instead of applying step changes on the real feeder. The simulation tests couple the modeled response of temperatures and flows in the feeder to modeled changes in the input. A special mathematical technique, Proper Orthogonal Decomposition (POD see [4, 5, 7, 6]) of temperature field predictions, is applied to further increase the speed of the modeling tasks and consequently the currently applied reduced models are very fast. This approach has two major advantages. Firstly, the tests do not need to be performed on the real feeder, preventing disturbance of the production process. Secondly, the results of these tests become easily available within a short period of time as GPS is very fast compared to the original CFD-simulations and less tests are needed to identify the response of temperatures and flows in the feeder to changes in the input compared to the conventional approach.
- A method has been developed to derive a kind of reverse response model to determine control input parameters enabling the achievement of constant temperatures and set points at the exit of the feeder even for the case of disturbances in the melt entering the feeder or disturbances in the feeder itself.

The resulting control model, that is derived from these tests using GPS, can be used for a large set of working points (e.g. a large range of loads) instead of for one single working point, as the response of the feeder to large variations in disturbances and process settings is determined. Consequently, the control model does not have to be rebuilt when a different working point for the feeder/furnace is selected due to e.g. the production of a different product (as long as the type of glass does not change). It is a fast way of setting up a complete and accurate control model without any risk for production. However, the reduced model is only applicable for one type of glass. Special mathematical techniques [5, 6, 7] are currently in development to enable model-based control for even larger operating windows and for controlling glass color changes.

The resulting control scheme for the industrial glass melt feeder is shown in figure 1. The temperatures in the feeder are controlled via the set-points of three PID controllers that adjust the fuel supply to the three heating zones in the feeder. These PID set-points are set by of the results of the MPC, which reads the values of the 9-grid thermocouple at the feeder exit and predicts the values of the input parameters to keep these thermocouple readings within a certain narrow range or to follow a pre-defined temperature course: based on the fast reduced model, describing the dynamic behavior of the feeder, the MPC determines the optimal values of the PID set-points such that the desired temperature (uniformity) at the feeder exit is attained. Next to the control objectives (desired temperature and temperature uniformity at the feeder exit), also several constraints are imposed to the MPC: the glass melt temperatures in the feeder may not exceed and drop below certain values; also the rate of fuel adaptation is constraint to avoid instabilities in the feeder. These constraints limit the flexibility of the feeder operation. Using the reverse response model, the optimal feeder settings (optimal PID set-points to ensure stable production at the desired glass melt temperature (uniformity)) can be determined easily.

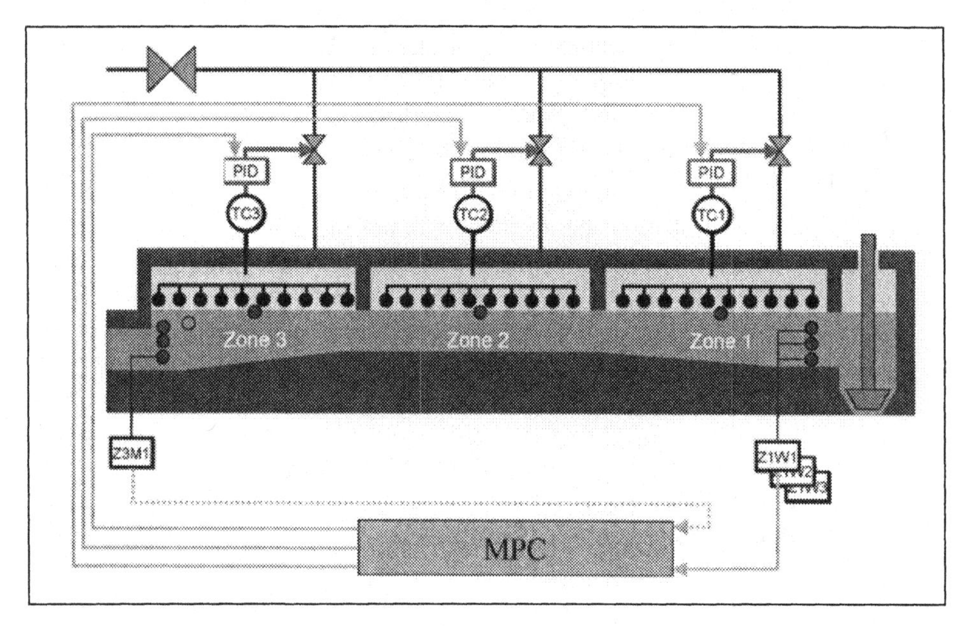

Figure 1. Control scheme for the industrial feeder. Blue arrows to PID controllers are the controlled settings of the PID controllers.

Although the feeder entrance temperature (uniformity) is measured continuously, in the here presented feeder control field test, this information has not been taken into account so far. Incorporation of the entrance temperatures in the MPC (indicated by the dotted line in figure 1) would allow the MPC to anticipate in the feeder (by adjusting the PID set-points) on temperature disturbances from the refiner. In the current feeder layout, an inline glass melt redox state sensor is positioned at the entrance of the feeder, acting as an early warning signal for redox disturbances in the glass-melting furnace (see figure 2). In the furnace to which the MPC controlled feeder is connected, up to 92 % foreign mixed recycling cullet is used acting as a source for redox variations. These redox disturbances may affect the temperatures in the feeder of the glass-melting furnace by changing radiation properties of the glass melt. Taking the signal of this redox signal into account in the MPC allows the feeder controller to compensate for the effect of redox changes on feeder temperatures.

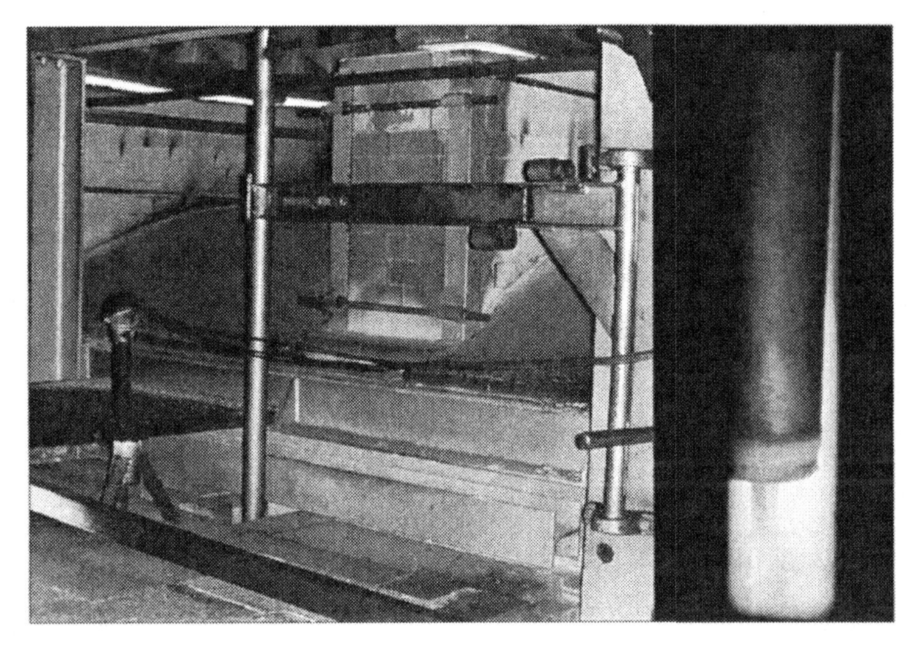

Figure 2. Inline feeder redox sensor positioned at the entrance of an emerald-green container glass producing glass furnace.

Preliminary Field Test of MPC Feeder Control

The application of the described MPC feeder control has been extensively tested (without redox state information and temperatures (uniformity) at the entrance of the feeder) on various production campaigns during the past months for a production feeder in emerald green container glass manufacturing. Figure 3 shows the impact of the controller on the average 9-point grid temperature, which is the main objective for the controller. In manual control mode, deviations in temperature in the 9-grid exceed +/– 2.5 degrees C, in some instances even more. It is clearly seen that the feeder temperatures become very stable (+/– 0.5 degrees C) once the controller is switched on. It should be noted, that the smallest change in temperature that is detected by the thermocouples is 0.2 degrees C, which makes the capabilities of the controller even clearer. This gives energy savings as well. Besides the increased stability, changes in set points are realized within a short period of time. Even automated transitions between largely different operating points (95 ton/day – 135 ton/day; different glass gob temperatures) have been performed successfully.

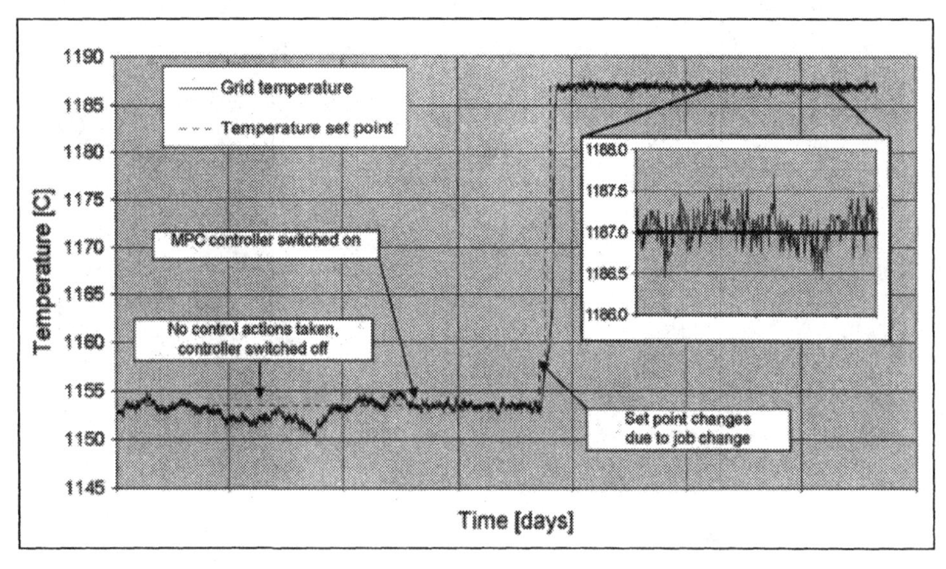

Figure 3. Result of controlling the average grid temperature in the feeder by making use of MPC control according to the approach as described in this paper.

Concluding Remarks

For an industrial emerald green container glass furnace, the application of MPC for temperature control in one of the feeders has been discussed. In this project MPC was based on a rigorous CFD model that provides detailed 3D information of the temperature distribution throughout the whole feeder. The POD based model reduction approach to speed up the calculation time of rigorous CFD models enables accurate model based control of critical process operating conditions without the need for expensive process testing and without any risk for production. Reduced model provides the same 3-D temperature profile information as the original CFD model within 0.2 oC. Simulation speed of the approximate model is at least 10.000 times faster than the CFD based simulations. Basis of MPC is the POD model reduction method. The POD model is derived from validated CFD models. MPC with POD reduced models has proven to be a solid track for model predictive control system designs for processes like glass furnaces or feeders.

The model predictive control system derived from detailed 3-D CFD models covers a large process area (set of possible furnace/feeder settings) instead of one working point. The paper showed that the MPC was very capable of controlling even large transitions or job changes with one model. For the feeder, an increase in temperature stability was shown, going from typical temperature variations of +/– 2.5 degrees C in manual control mode to +/– 0.5 degrees C during MPC. Even more stable production could be achieved taken in the MPC into account also information from:

- The glass melt temperatures at the feeder entrance,
- The redox state (variations) at the feeder entrance measured by an inline redox sensor, and
- Electrical boosting as additional manipulative value to compensate for differences between left and right of the feeder.

ACKNOWLEDGEMENT

The development of the approach that is described in this paper has been supported by EET (Economy, Ecology and Technology), a program of the Dutch government. The authors would also like to thank Sven-Roger Kahl, the manager furnaces of REXAM in Dongen, for his contributions to the project.

REFERENCES

1. Krause, D., Loch, H. (Eds.): "Mathematical simulation in glass technology", Schott series on glass and glass ceramics, Springer, Berlin, Heidelberg, 2002.
2. Op den Camp, O. and Verheijen, O.: "Application of fast dynamic process simulation to support glass furnace operation", 64th Conference on Glass Problems, October 28–29, 2003, Urbana-Champaign, Illinois.
3. Op den Camp, O., Verheijen, O., Huisman, L., van Deelen, S., and Backx, T.: "The use of glass process simulation for control of glass quality in glass production", 20th International Congress on Glass, September 27 – October 1, 2004, Kyoto, Japan.
4. Astrid, P., Huisman, L., Weiland, S. and Backx, A.: "Reduction and Predictive Control Design for a Computational Fluid Dynamics Model". 41st IEEE Conference on Decision and Control, Las Vegas, December 2002.
5. Astrid. P., Weiland, S., and Twerda, A.: "Reduced order modeling of an industrial feeder model", Proceedings of IFAC Symposium on System Identification, Rotterdam, The Netherlands, 2003.
6. Astrid, P., Weiland, S, and Willcox, K.: "On the acceleration of a POD-based model reduction technique", Proceedings of the 16th Symposium on Mathematical Theory, Network and System, Leuven, Belgium, 2004.
7. Astrid, P., Weiland, S, Willcox, K, and Backx, T.: "Missing point estimation in models described by proper orthogonal decomposition", Proceedings of the 43rd IEEE Conference on Decision and Control, Paradise Island, Bahama, December 2004

Application of IR-Sensors in Container Glass Forming Process

Joop Dalstra, XPAR Vision B.V., The Netherlands

Abstract

This paper describes the functionality of equipment measuring the infrared (IR) radiation emitted from each single glass container directly after the forming processes in glass production plants. The IR images from the hot containers show the glass distribution, cooling behavior of the article, but also glass faults in an early stage after forming. Combination of this technique with statistical methods enables process control and the IR imaging method supports operators to improve the settings of the forming machines or gobbing process and monitors the performance of the swabbing process. Examples are shown for the application of these IR imaging sensors in an industrial environment for Hot End Quality Inspection, Process Monitoring, Process Control and Process Optimization.

Introduction

The bottleneck of the container glass industry: the lack of process information.

The container glass process can be divided into two time zones: the hot end (melting & forming) and the cold end (annealing & inspection). Newly formed containers must be annealed in the annealing lehr. The annealing time is dependent on the type of product and lasts about one hour or more.

For this reason it takes at least 1 hour production time before 'wrong' containers are discovered. Only then the warned hot end machine operator is able to take remedial action. He is forced to rely strongly on his own skills and knowledge, since the relation between the product quality and the hot end process parameters is mostly unclear and mostly too late.

The machine operator will need another hour to be sure of the effectiveness of his intervention. In most cases the hot end adjustments turn out to be ineffective or even worse, they often show reverse effects. Then, as a consequence at least two hours production time are lost! Therefore a system and methodology is needed which is capable of performing quality inspection tasks and simultaneously monitoring the hot end process performance in real-time.

Instrumental Method

The basic principles of infrared technology. Hot end containers, freshly formed, radiate infrared radiation. This infrared radiation is dependent on the local temperature of the glass and the local thickness of the container. For instance, when the temperature of the glass increases or if the sidewall becomes thicker, more infrared radiation is emitted.

The XPAR infrared camera system captures images of containers immediately behind the IS-machine (figure 1) and before the coating funnel. It needs one meter distance from the conveyor belt.

Figure 1. The infrared camera (cabinet at right hand side: X) is placed immediately behind the IS-machine and before the coating funnel.

It appears that, from this position, an infrared image of a newly formed container contains far more process and product-related information than a visual light CCD camera could provide for. This so-called thermal image is an accurate representation of the heat and glass distribution of the whole container (see figure 2). The XPAR infrared camera combines excellent sensitivity with sophisticated optics. It generates high-resolution images of hot containers: changes in glass thickness of 0.1 mm can easily be distinguished.

Being processed and analyzed by the system, the IR-image provides relevant and real-time information that could help to control and to optimize the glass forming process as well as the product quality. Contrary to hot end process information cold end information is always too late. This explains why the information from cold-end inspection equipment is disqualified for process control and optimization (too late and incomplete and labor intensive).

Applications of Infrared Technology

What are the possibilities for a system, which provides information about the glass distribution and the temperature distribution of the whole container? Quite a few: At the moment all the applications of infrared technology in the glass industry can be divided into four categories:

- Quality Inspection;
- Process Monitoring;
- Process Control;
- Process Optimization.

For each category an example of an application is presented in the following sections.

Application of the IR Imaging System

Quality Inspection

Defects. The infrared technology of XPAR Vision detects each small disturbance in the glass material caused by a defect, such as a stone, a blister, a thin spot, or adhering glass particles. In figure 3 you see an infrared image of a bottle containing a stone is shown. A 3D representation (showing the IR emission along the surface area of the bottle) of the infrared intensity is also shown in the same figure. Clearly one can see the huge impact of the defect on the infrared radiation. By means of infrared technology it is therefore possible to detect and reject critical defects.

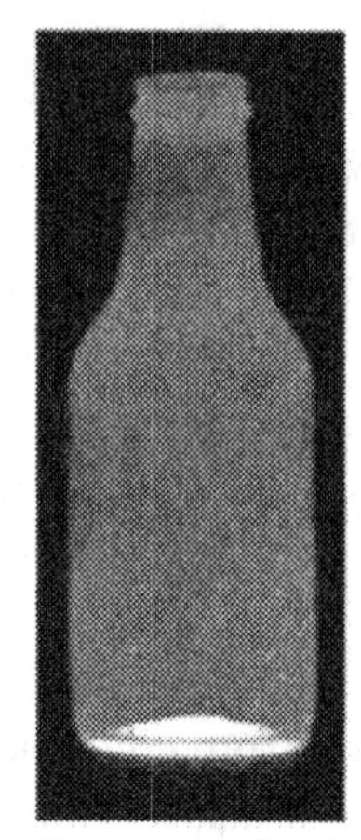

Figure 2. An infrared image of a beer bottle. Yellow (light) means high infrared radiation levels, blue (dark) means low radiation levels.

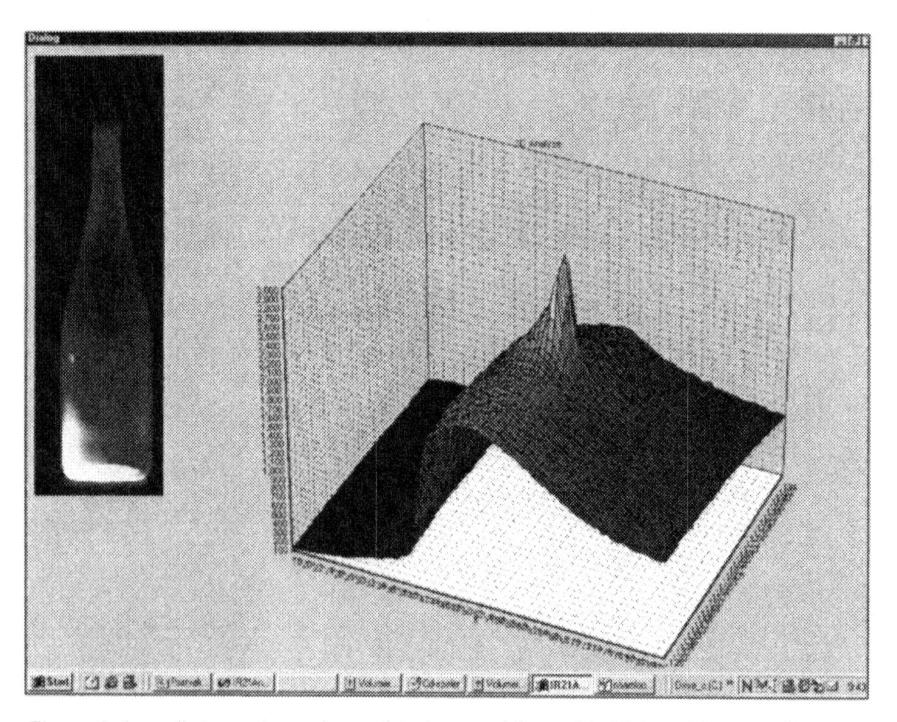

Figure 3. A small stone gives a large disturbance of the emitted infrared image pattern.

Thin ware. Thin bottoms are a major problem for the glass producers. Cold end Side Wall detection systems do not detect these thin bottoms. With infrared technology it is very easy to detect thin bottoms. The radiation of a thin bottom is far less than the radiation of a normal bottom. But if the bottom is colder and not thin at all, the infrared radiation is also less. Question now is: in which way can one distinguish between a thin bottom and a cold bottom, or between a hot bottom and a thick bottom? XPAR Vision developed a method to enable this distinction.

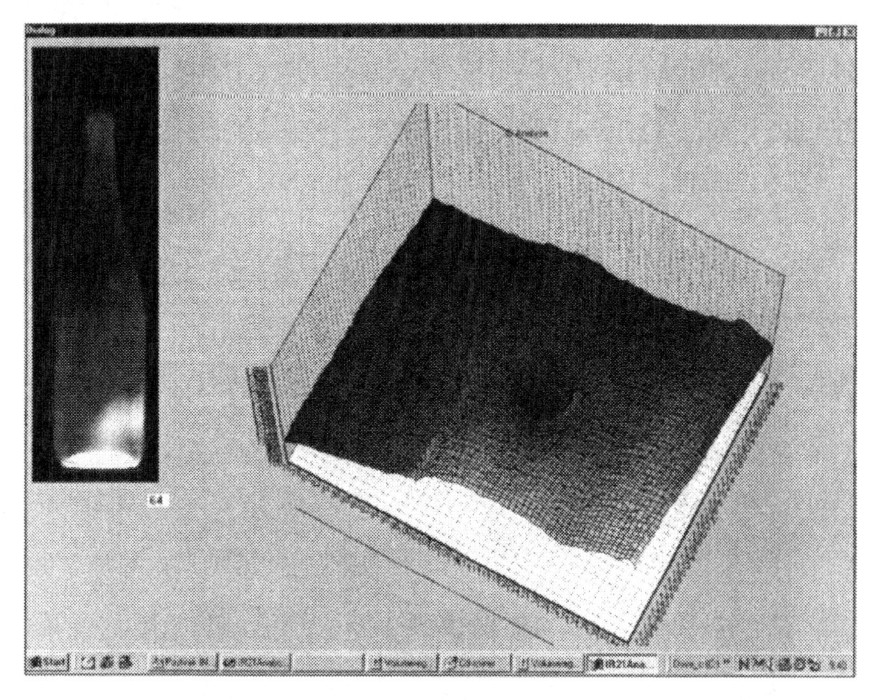

Figure 4. A blister resembles a hole in the infrared images.

The amount of radiation is dependent on the amount of glass of the bottle. If for some reason the bottom thickness decreases, then the amount of glass in the bottom area is also less. Because the total amount of glass must be constant, the missing glass must be somewhere else. It appears that the 'missing' glass has shifted to the shoulder area of a bottle. More glass in the shoulder area means the infrared intensity of this area will also be increased. So a combination of a decreased infrared intensity of the bottom area (caused by less glass) will be accompanied by an increased infrared intensity of the shoulder area (figure 5). In this way you can detect the thin bottoms with 100% efficiency. Also thin or choked necks can easily be detected at the Hot End location.

Another example of inspection is the detection of 'thin necks' (less glass in the neck area).

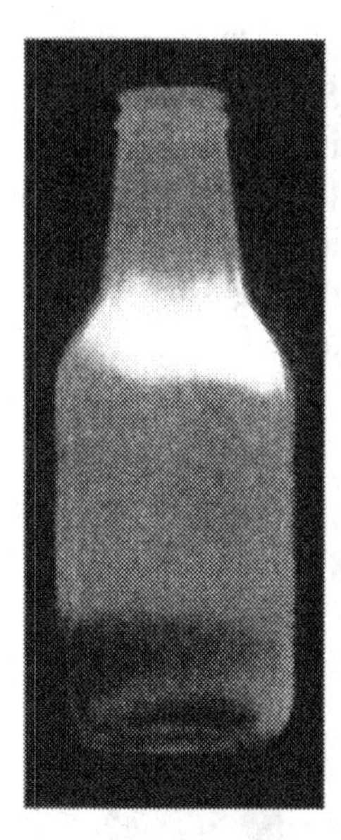

Figure 5. A thin bottom means a shoulder area with more glass!

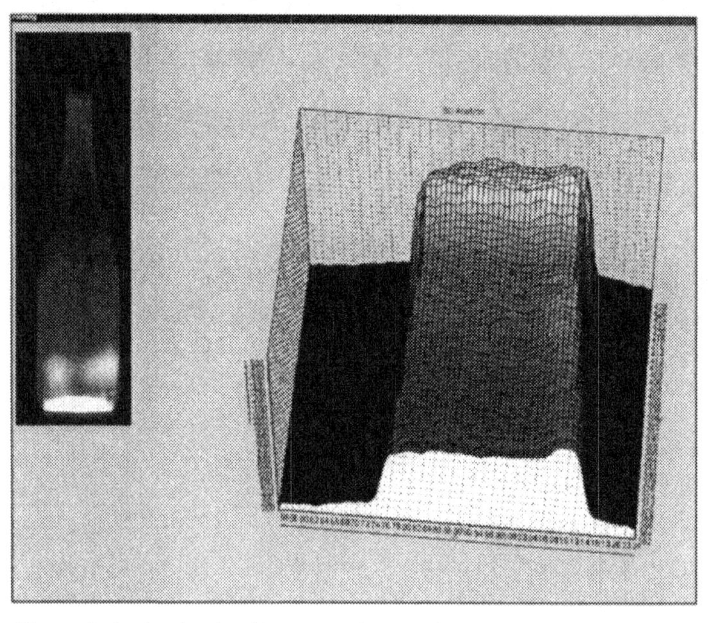

Figure 6. A wine bottle with a normal glass distribution in the neck area. The Y-axis of the graph represents the emitted IR-intensity. The emitted IR-radiation is evenly distributed which means the glass thickness is also constant.

In figure 6 you see a bottle with a normal neck thickness, a 3D-presentation of the emitted IR-radiation of the neck and the top of a wine-bottle. The Y-axis represents the amount of radiation. In figure 7 a bottle is shown with a thin neck. As you can see the emitted IR-radiation is much less then in figure 6. This shows the large sensitivity of infrared technology with regard to the glass distribution.

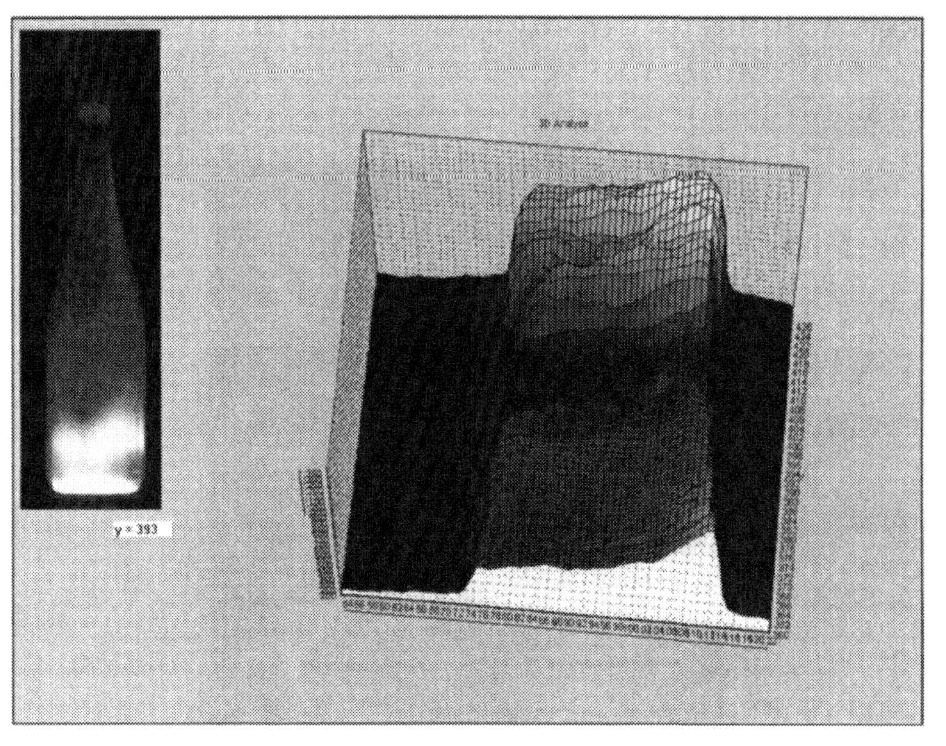

Figure 7. A bottle with a very thin neck. The neck area emitted very less IR-radiation. This means that the wall of the neck area is very thin.

Shape and verticality. The infrared system of XPAR Vision is capable of measuring the shape and the verticality of a bottle or jar. Even the smallest deviation in the shape is notified straight away. Glass makers can therefore identify and detect any problems at an early stage of the production process, before many other containers are manufactured with the same wrong shape. Sometimes the position of the top of a bottle ('shifted finish') is offset from the ideal center line, caused by a malfunction of the glass forming process. These critical defects are identified and rejected at the hot end (figure 8).

Figure 8. Detecting and rejecting bottles with a poor shape at the Hot End has many advantages! The right hand sided bottle is an example of a shifted finish.

Process Monitoring

The most obvious and evident advantage of this new infrared technology is the automated real-time monitoring of glass distribution and wall thickness of the whole container per each individual container.. The real-time information of the glass distribution of each station is vital for a good process control and prevention of faulty products.

An infrared camera is a non-contact sensor that registers the emitted infrared radiation of the whole container, even at the difficult inspection areas at the neck and at the bottom. Because the infrared system is synchronized with the IS-machine the performance of each mold and station can be monitored.

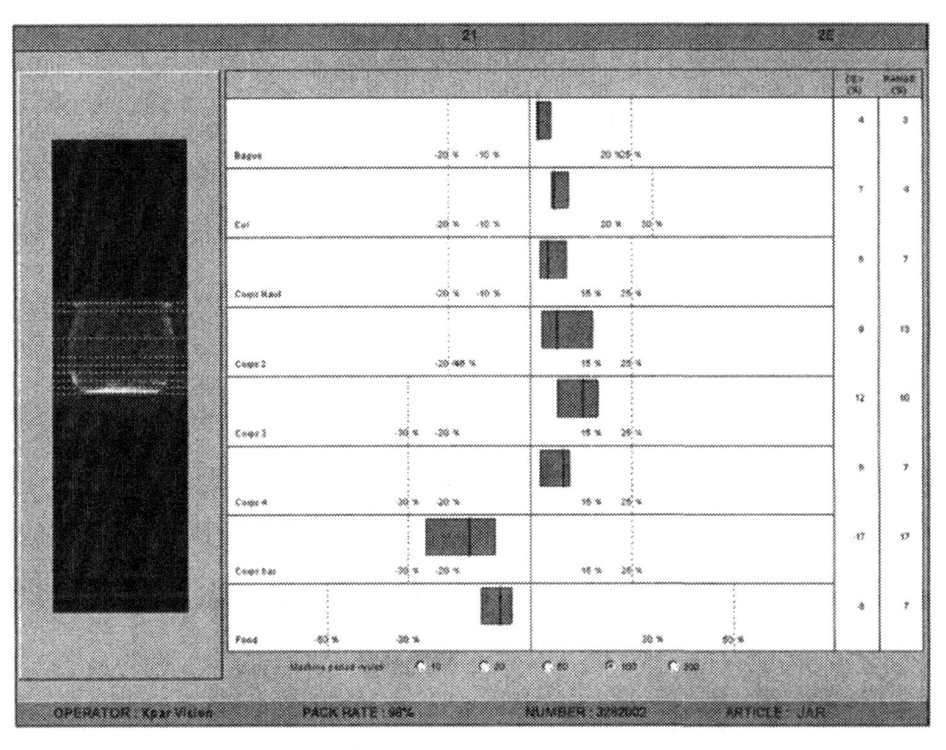

Figure 9. The IR-camera system gives the Hot End operator real time information about the glass distribution. The IR image of the jar is divided in several (here eight) measuring zones (white lines). For each zone an optimal set point is calculated by the IR-camera system. The squares in the right graph gives the deviation of the set point and the variation of each measuring zone.

Deviations in infrared intensities of each product are constantly measured, registered and compared to the stored image of the ideal infrared image or so-called "infrared blueprint". This is a valuable reference that always corresponds to a good-quality-product. Of course there are blueprints for each product and each mold. XPAR Vision has developed a verification method for finding of the right glass distribution. The method is based on the construction of the so called IS-machine chart for each product. The method enables real-time verification of the glass distribution of each product while passing the camera!

Process Control

The amount of process information stored in the infrared images is big and very useful for control of the glass forming process. Some information can be used for automatic controlling of the glass forming process. One example is the new development of the Infrared Gob weight Control (XPAR IGC) of XPAR Vision. It controls the weight of the gob by means of information from the infrared radiation. The amount of infrared is related to the amount of glass, this is of course related to the weight of the bottle. In this way XPAR Vision can keep the weight of the gob constant in time. In figure 10 you can see the setup of the XPAR-IGC system.

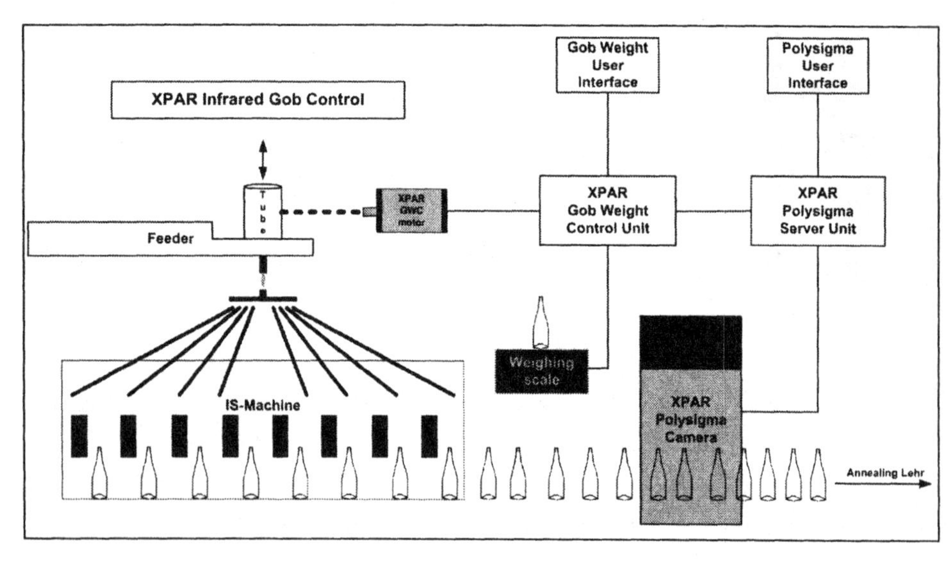

Figure 10. A gob weight control system. The Polysigma is the IR-camera system. The XPAR Polysigma Server Unit is a central computer for the data storage and for sharing the information for multiple users.

In the future also other glass forming processes, like the time settings and the cooling of the IS-machine can be automatically controlled by means of infrared technology. In this way the "human factor" will be minimized in the future.

The Polysigma User Interface is a computer for displaying the information to the Hot End operator. The Gob Weight Control Unit contains the motor controllers for adjusting the tube height. The Gob Weight User Interface presents information about the IGC system.

Process Optimization

The real-time process information makes it possible to optimize the glass forming process. The first example is the optimization of the glass forming process by optimization of the gob loading in the blank mold. The loading of the gob in the blank mold is very important for the final quality of a bottle. A small deviation in the loading position of the gob in the blank mold decreases the quality of the bottle considerably and is also responsible for a lot of problems in the glass forming process. Adjusting the positioning of the gob in the blanc mold is nowadays still dependent on the craftsman skills of the hot end operator. In practice this mostly means that the gob loading is far off it's the optimal settings and the quality and efficiency of production process is decreased.

With the infrared technology of XPAR Vision it is possible to optimize the position of the falling gob in the blank mold. Please refer to figure 11. It displays the User Interface of the XPAR Polysigma.camera system. The graph represents the infrared emissions of the various zones of the infrared image of a beer bottle. The infrared intensity of the bottom area is very noisy (lowest graph). This is the result of a "bad" loading. Every bottle has a different bottom thickness and a different glass distribution. At the end of the graph the optimal loading of the gob has been found. The noise of the infrared radiation is gone and the glass forming process has now become stable and the pack-to-melt ratio has been increased with several percents!

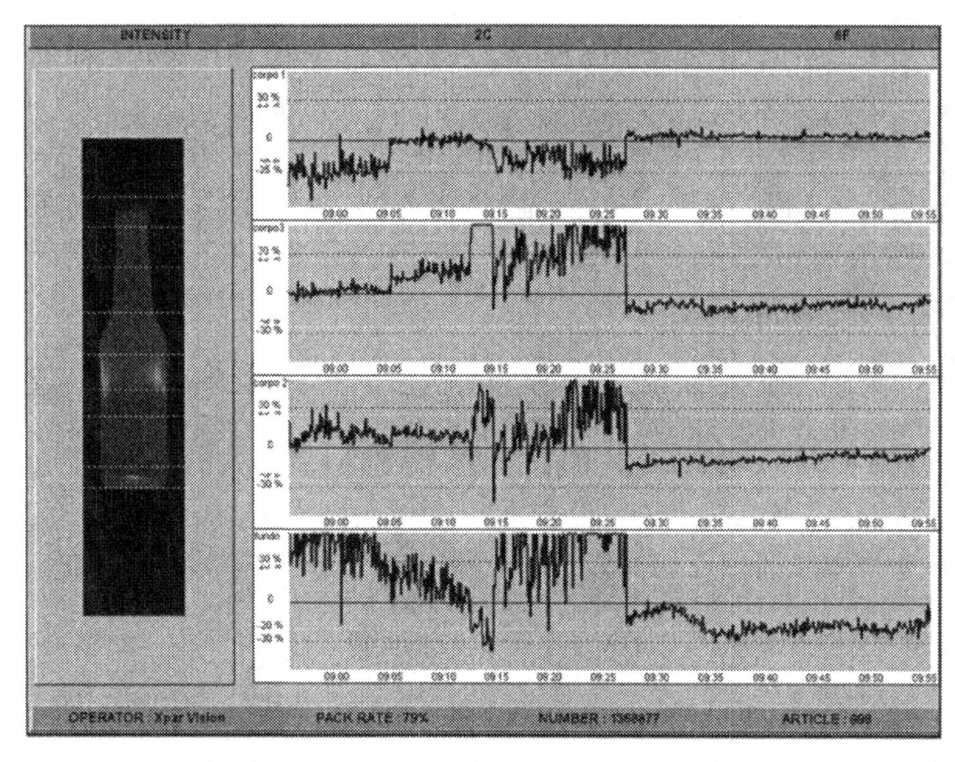

Figure 11. Bad loading of the gob in the blank mold is the cause of many problems in the glass forming process. The graphs shows the IR emissions of each zone in due time. The Y-axis presents the amount of radiation relative to the set point (percents). The X-axis is the past time. The graph starts at the right (present) and goes to the left (past). The white dashed lines in the IR-image (right) marks the different zones.

Another example is swabbing. The molds have to be swabbed with graphite, otherwise the glass will stick to the wall of the mold. But swabbing is also a very big disturbance for the glass forming process. Immediately after the swabbing the glass distribution is very bad and the bottles must be rejected for a several cycles. With the infrared system one can monitor the effect of a swabbing. See figure 12. This is also a image of the User Interface of the Polysigma infrared system. The influence of the swabbing is very noticeable. The infrared intensity of the bottom area rises suddenly and at the same time the intensity of the shoulder decreases for several dozens of cycles.

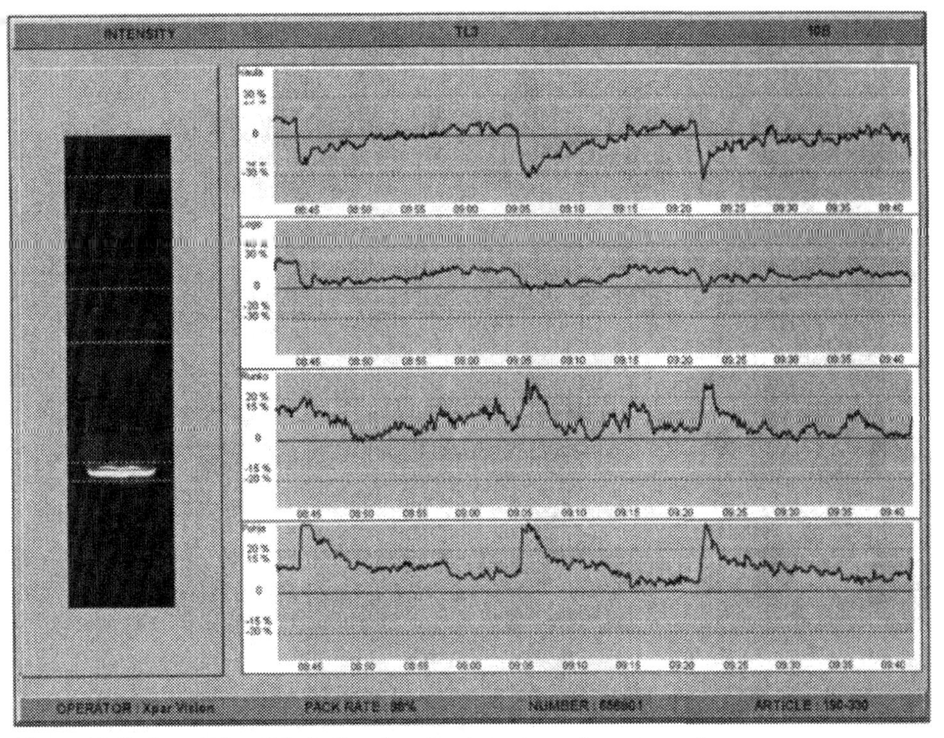

Figure 12. Bad swabbing disturbs the glass forming process for many cycles.

This means that the bottom area got thicker and the shoulder area got thinner. With this infrared information you can now optimize the swabbing. The deviation of the infrared intensity by swabbing is directly related to the way the operator performed his swabbing task. Less disturbance of the infrared intensities means a better swab result with a minimum of disturbance of the glass process and the quality of the bottle/jar. In this way the operator can be trained to optimize his swabbing routine and have less negative influence on the quality and performance of the glass forming process. Compare with figure 13. This is the swabbing result of a Japanese container glass process. The influence of the swabbing is minimal.

Conclusions

The proven benefits of the XPAR infrared system. The XPAR infrared system, thanks to its strategic position in the production process, provides instant process information to the IS-machine operator and helps him to receive immediate feed back on his process interventions. Upon installation of the XPAR infrared system, the user can stabilize the process considerably without much effort. The most evident direct related benefit will be a structural increase of the pack-to-melt ratio.

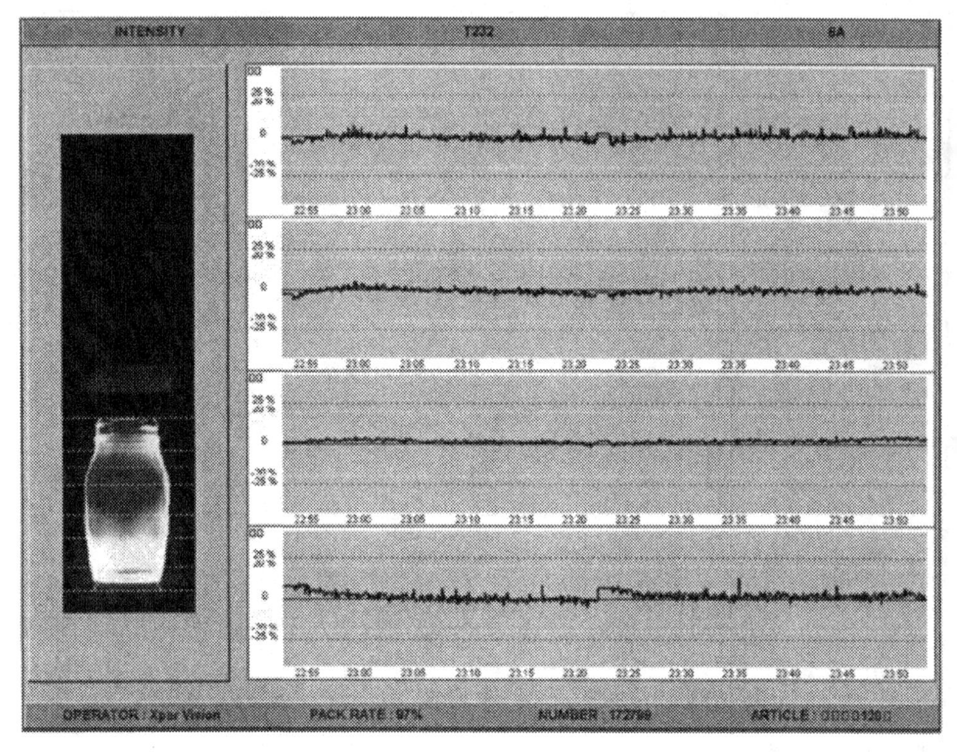

Figure 13. A good swabbing method means less disturbance of the glass forming process. Minimal disturbance of the glass forming process means less faulty ware and a higher quality.

A troubling mold or station of the IS-machine is automatically tracked and immediately tracked, since the XPAR infrared system is synchronized to the IS-machine.

By only using one single camera system the glass distribution of the whole container can be monitored. This simplifies the detection of thin spots even at the difficult accessible inspection areas of a glass container.

Other critical defects such as stones, blisters, bird swings etc. can also be detected and removed. But the most unknown benefit of infrared process monitoring is the tremendous potential to gain more knowledge and insight in the forming and gob forming process, which results in a higher pack-to-melt ratio.

Transmitted and Reflected Distortion of Float and Laminated Glass

Ulrich Pingel, ISRA GLASS VISION GmbH and
Peter Ackroyd, ISRA VISION SYSTEMS Inc.

Preface

Optical distortion is an important quality parameter in float glass production and fabrication. Transmitted and reflected distortion individually as well as their relationship in combination need to be understood and monitored to guarantee an acceptable product quality.

This presentation explains the physical phenomenon, the adaptation of measurement techniques and parameters which can be important in determining final product quality.

Optical Distortion in Glass

Distortion is known as an optical manipulation of an image in transmission or reflection. Distortion is visible nearly everywhere, e.g. lenses or glasses use this effect as well as a concave mirror in the bath room or the distorted reflection in the glass-front of a building.

The quality requirements of optical distortion in plane glass are becoming more critical. Windscreens are mounted at much lower installation angles, back-projection screens use mirrors in more critical installation conditions, thinner glass, bending becomes more complex, surfaces larger and customers are requiring better quality.

The existing float process is approaching its technical limits and may only be improved long-term and at enormous cost.

Hopeless?

One solution is to understand the effects of distortion much better, the sources of distortion in the manufacturing process and to continuously monitor those effects through precision measuring techniques and statistical process control.

As in many cases, the relevant wavelength and tolerance of distortion either in transmission or in reflection is a function of geometry. The human eye located nearby the glass will be more sensitive for narrow distortion. On the other hand long wave distortion will be easily detected far away from the glass.

Optical distortion can be demonstrated by superimposition of 3 basic effects:

- Thickness variations create spherical or cylindrical lens effects
- Change of refractive index, e.g., due to inhomogeneity or ream creating lens effects
- The undulation of glass (2 parallel surfaces) creates reflected distortion

Real glass may contain all 3 kinds of these effects.

Figure 1. Reflected distortion in a building facade.

Figure 2. This picture is taken viewing through the windscreen of a car. The effects of the distorted car body lines are caused by the transmitted distortion of the windscreen.

Figure 3. Reflected distortion of bent glass.

How to Measure

For statistical process control, it is necessary to measure the distortion in order to be able to compare different production periods and lines. The measurement should represent all important quality parameters, further more it should have linear and continuous behaviour. An automatic measurement will support reliable SPC, quality control and quality selection.

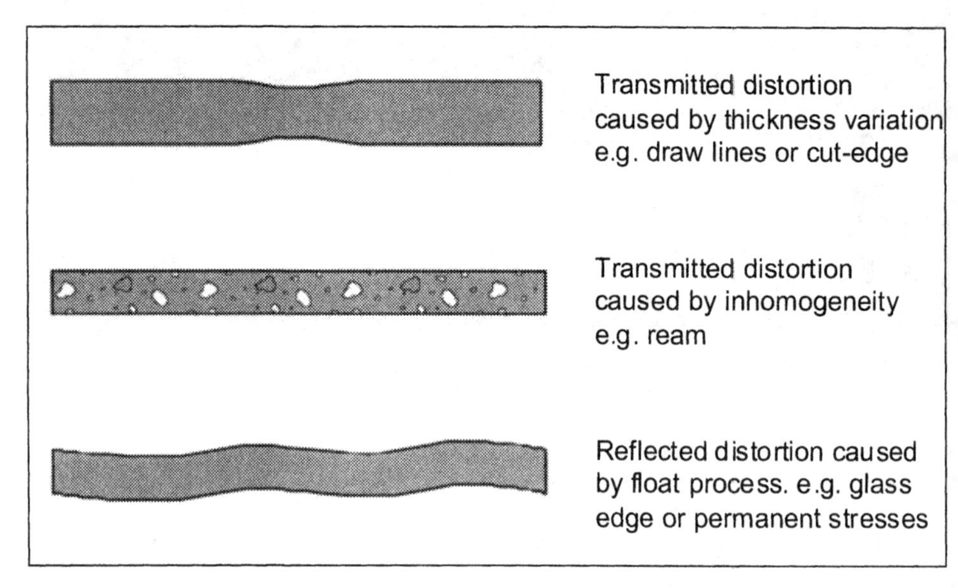

Figure 4. The 3 basic distortion effects.

In contrast to the nominal measurement of "ZEBRA-ANGLE", the measurement by the unit "DIOPTER" [2] gives a physically based value widely used in the glass industry because it is quantifiable and repeatable. It enables an accurate measurement without limits and may be used in automatic inspection systems.

It is necessary to analyze the glass distortion at different wavelengths. For instance: Chemical ream may generate distortion at higher frequencies than draw lines during the float process. This often enables the determination of causes of different distortion problems for instance due to the melting-, float bath- or annealing lehr processes.

The Online Moiré Technology is based on an interference of two grids. A glass with distortion changes the interference pattern which is captured by modern linear line scan cameras. Thanks to the interference method, the measurement is not effected by environmental light conditions, glass color or glass thickness. The technology may be used to inspect for distortion and other glass defects in transmission and reflection with highest accuracy and spatial resolution. Different low- and high pass filters, adapted to typical distortion wavelengths enable a continuous process and quality monitoring online or offline.

Figure 5. Moiré interference of two grids.

Transmitted Distortion

Transmitted distortion may be viewed by the flickering effect looking through the glass, either perpendicular or at a specific angle. Chemical inhomogeneity as well as thickness changes cause optical distortion at different wavelengths. Main sources of transmitted distortion e.g. ream, draw lines, are caused by the melting and float bath process. Laminated glass may include distortion of different kinds and grades due to the lamination process.

A special effect occurs if bent glass is viewed at a large angle. This is explained further on p. 32.

A favorable way to measure and analyse transmitted distortion is the metric unit Diopter or Millidiopter. It describes the optical power of distortion as a lens effect [2] and is inversely proportional to the focal length.

Because the unit "Millidiopter" is an independent SI measurement of the distortion, distorting effects with short wavelengths like ream up to wide distortion profiles like roller wave distortion may be measured and graded.

Reflected Distortion

Reflected distortion is generated by the shape of the glass surface. This may be created by undulation (e.g. roller wave of tempering furnace, float distortion), stress or even by designed curvature.

Reflected distortion is easily visible on surfaces which are far away, and large mirrored targets at a long distance to the mirror.

Important reflected distortions are the ones created in the float process or the longer wave distortion due to thermal processing and residual stress.

Also a nominal curvature includes reflected distortion, so in some cases it is important to separate between required and unwanted distortion. A good example is automotive glass. Designed shape, bending imperfections, float distortion

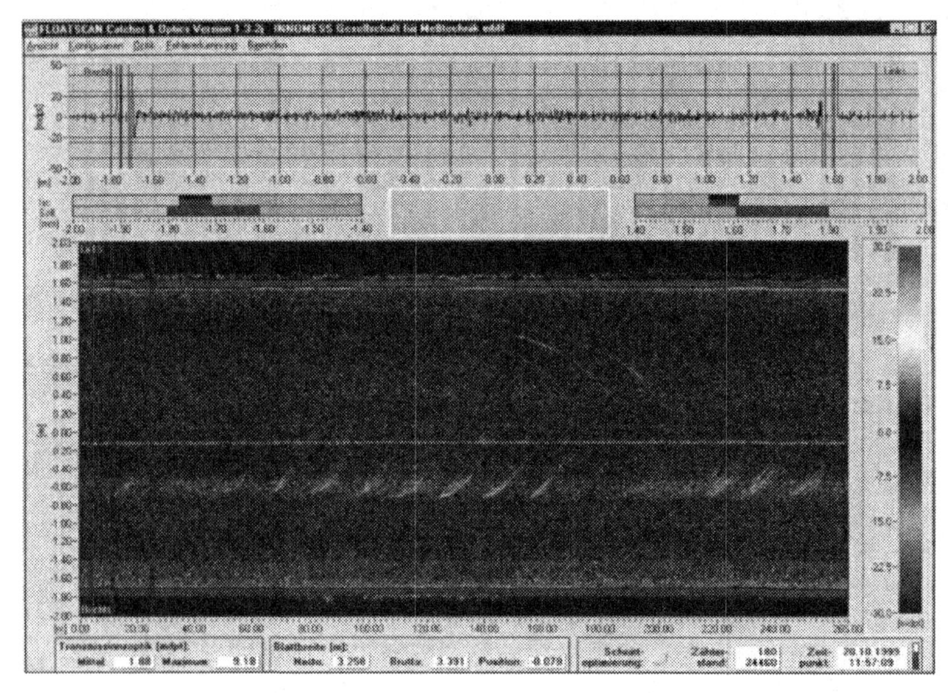

Figure 6. Transmitted distortion of float-glass ribbon of 265 m length as a color-coded image. The marked distortion is caused by glass inhomogeneity. The distortion is coded by color with a range of –30 mdpt (blue)to +30 mdpt (red)).

Figure 7. Transmitted distortion caused by a small glass defect.

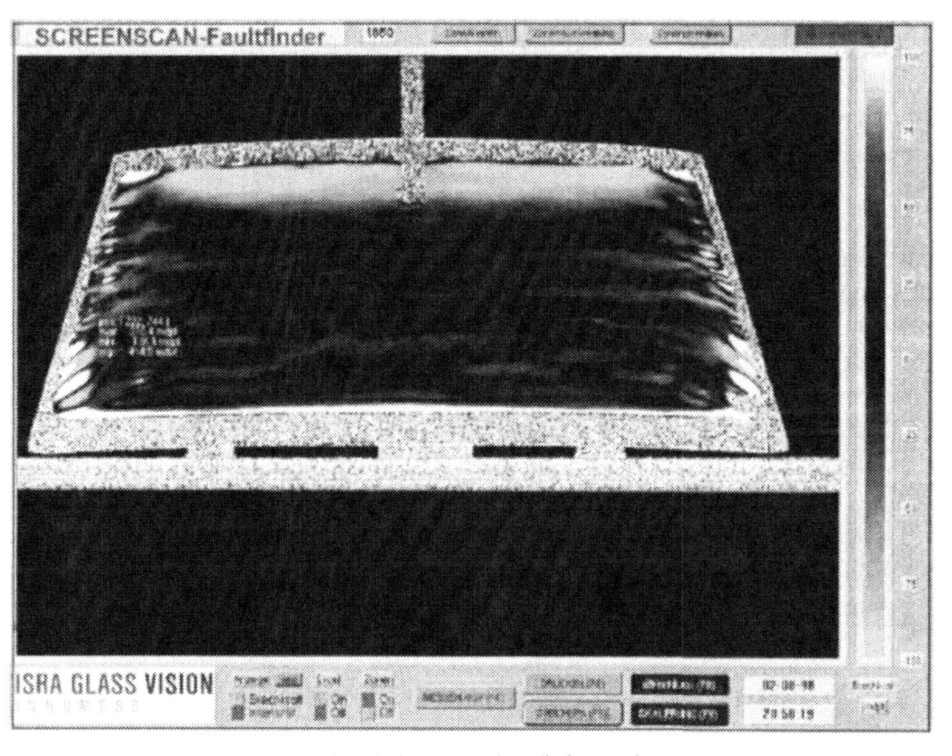

Figure 8. Transmitted Distortion of a windscreen at installation angle.

and lamination problems take part in the final shape. Line and area distortion, with longer or shorter wavelength characteristics may be observed.

Special care is necessary with reflected distortion on float glass used for automotive, facade or mirror applications. The reflected distortion may be described by its radius, or, in a better way, by its optical power, measured in the unit Diopter or Millidiopter.

The ability of the Online-Moiré-Technology to measure, analyze and grade the distortion in the unit Millidiopter leads to a repeatable and long-term comparable measurement, online and offline. So reflected as well as transmitted distortion may be judged using the same quality grading.

Reflected Distortion Results in Transmitted Distortion:

Due to lamination. There is an important effect, especially on laminated glass, because the reflected distortion of both outer surfaces will strongly determine the transmitted distortion of the laminate. The first figure shows the reflected distortion of the front–side of a laminate. The second figure is the transmitted distortion of the sample. In this test, the reflected distortion of the back-side of the laminate was nearly zero.

Due to the lamination process, the inner surfaces of the two plies are optically connected, so only the outer surfaces are responsible for the thickness variation and therefore the transmitted distortion of the laminate. The stiffness of the glass, amplitude and radius of the reflected distortion will influence the transmitted distortion.

If you compare figures 9 and 10, you will see, that significant negative or concave (green / blue) transmitted distortion may be observed where the reflected distortion shows strong convex (red/yellow) distortion lines. Some significant lines are marked with A and B. As a result of this test, the reflected distortions influences the transmitted one with inversed polarity. This should be taken in consideration as soon as glass is laminated

Due to viewing angle. Glass undulation shows a very little effect of transmission distortion if viewed perpendicular through the glass. This may change rapidly if you rotate the glass. This effect is widely used in the ZEBRA-test, but it also complicates the design and manufacturing of bent glass, but also flat glass applications are effected. The following comparison shows the same windscreen measured perpendicular and at 70 degree tilt angle. The effect is so strong, the scaling also had to be changed from +/- 20 mdpt to +/- 220 mdpt.

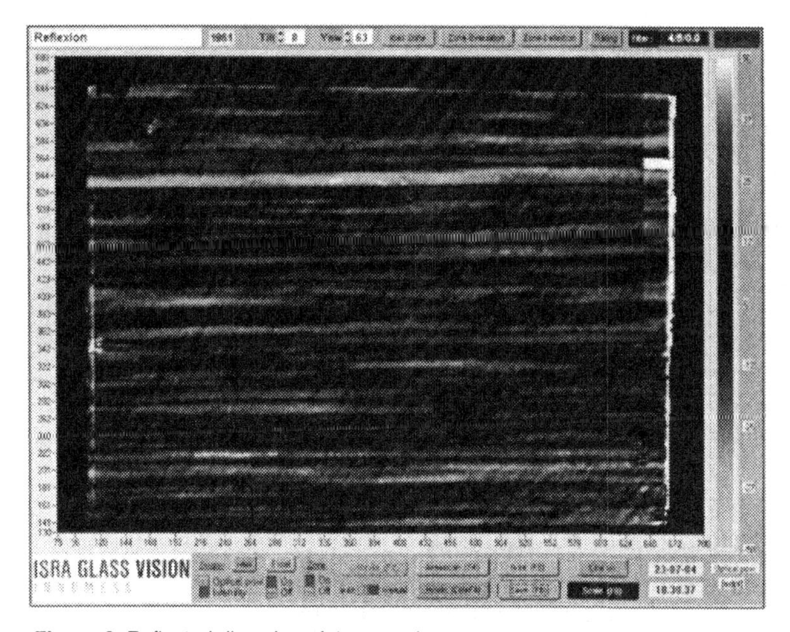

Figure 9. Reflected distortion of the sample.

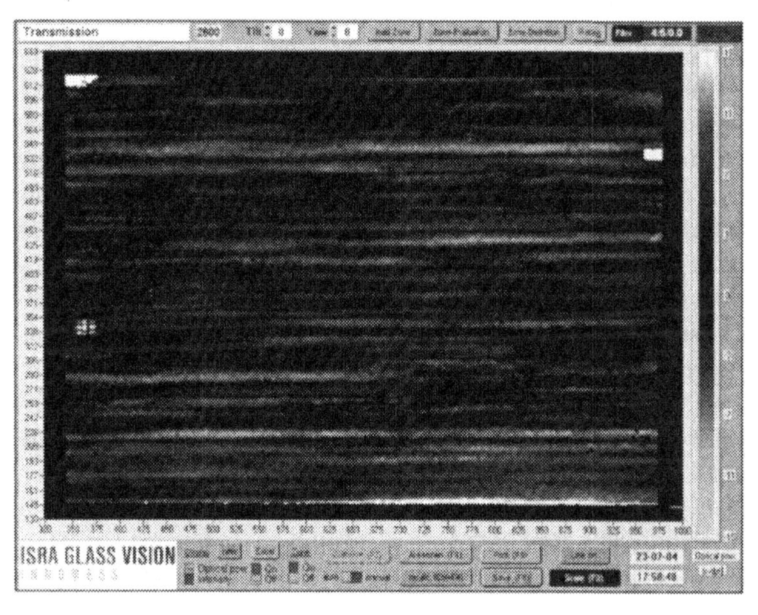

Figure 10. Transmitted distortion of the laminated sample.

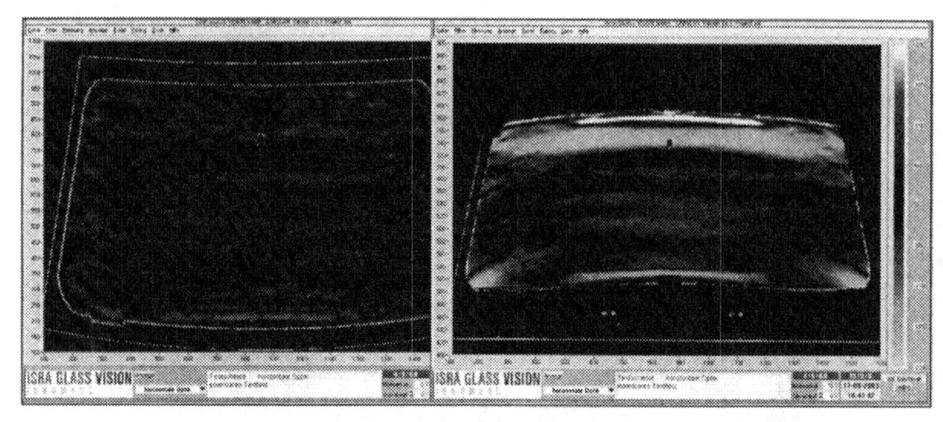

Figure 11. Transmitted distortion of a windscreen at perpendicular (left) and at 70 degree (right).

How to Measure in Production

Distortion, whether reflected or transmitted may be measured online or offline by means of the Online-Moiré Technology. Because of the random occurrence of ream distortion, it is most common to inspect transmitted distortion online. Reflected distortion may be inspected online or offline depending on the individual requirements.

A typical offline measurement can be made with the LABSCAN-Sample. This product measures cut samples of flat glass up to 40" x 50" and scans for point defects, transmitted and reflected distortion. Its spatial resolution of less than 0.2 mm in combination with ISRA`s proprietary Online-Moiré-Technology, allows detection of the distorting halo of small glass defects and also large area linear and spherical distortion. The typical usage is to analyze float and fabrication processes for development, quality control and for automatic sample testing.

Figure 12 shows a FLOATSCAN-Catcher&Optics in a typical online application on a float-line. The whole ribbon is automatically inspected for point defects and transmitted distortion. The result is analyzed and graded according to the customer's requirements and then input to the cutting optimization computer. The wavelengths of the distortion may be analyzed differently for ream and draw-lines, allowing different tolerances and actions.

Optionally, the reflected distortion channel may be added to measure the glass undulation online. The system features very high accuracy and spatial resolution as well as comprehensive quality analysis and connection to the cutting optimization computer.

Figure 12. Online inspection system for float glass.

Conclusion

Transmitted as well as reflected distortion are significant quality factors in glass production and fabrication.

The Online-Moiré-Technology allows the accurate and precise measurement of transmitted and reflected distortion in the SI unit Diopter and to lead to a closer understanding and new ways of process and quality improvement.

REFERENCES

1. Moiré Technology
   http://www.optics.arizona.edu/jcwyant/Optics513/Lab/MoireTechniquesLab12.pdf
2. SI Unit Diopter
   http://galileo.phys.virginia.edu/classes/202.stt.spring04/Thornton%20Lecture%20Notes/Lecture%2016.3-22-04.pdf

Inspection: Going Beyond Just Finding Defects

Christian von Ah, Emhart Glass SA, Switzerland

Abstract

Rapid changes in technologies and glassmaking processes have transformed the way manufacturers see inspection. Inspection devices now must run at increasingly faster line speeds and be more flexible. They must provide a wide array of information that can be used to optimize the process. This paper will cover the development of sophisticated vision inspection and servo technologies and how they have served as catalysts for the development of new, fully integrated systems that both inspect finished ware and provide valuable information that help improve the process. The paper also will show how Emhart Glass has designed a whole new family of inspection machines covering every aspect of finished ware inspection: the Veritas iM performing plug, ring, and dip gauging as well as check inspection and non-contact wall thickness measurement; the Veritas iB performing sealing surface, base, base stress, vision plug, and vision dip inspection; the Veritas iC looking for sidewall defects and sidewall stress, as well as giving information on dimensional measurements, such as body diameter, height, bent neck, and lean.

Introduction

It is well known that glass—as a primary packaging material—is facing difficult times. Today, glassmakers must be innovative and flexible in order to compete with glass alternatives. The challenge is to react quickly to a demanding and changing market, increase container quality, and, at the same time, stay productive.

One element that is necessary to ensure a good quality of packed ware is the inspection machinery. Sophisticated inspection technologies are required to meet these challenges. The time has passed when one bottle shape was adequate for all. Today, every wine, beer, soda, or liquor bottle needs its own characteristic size and shape.

- Inspection machines must be able to adapt to any shape of container.
- Every inspection needs to be fast, reliable and repeatable.

To accomplish this mission, state-of-the-art technology, such as high-resolution cameras, servo and stepper motors, and advanced software, are required.

This paper will discuss what technologies can be used and their underlying benefits. Furthermore, it will describe what else is important in the inspection business and in what areas inspection machines can do more than just inspect.

Today, Emhart Glass meets these new and ever-growing demands through a completely new inspection platform called Veritas. Veritas is a complete line of inspection machines that incorporates some of the best inspection and handling technologies in simple, easy- to-use machines.

Flexibility Drives Machine Design

There were times when it was possible to make one beer bottle for all beers and one or two wine bottles for all wines. However, the brewers and wine makers, as well as many other users of glass containers, discovered that the shape of a container could distinguish them from their competitors. Therefore, they demanded individual shapes for every wine, beer, and water bottle.

What does this mean for the glass making industry? Shorter production cycles and, therefore, more job changes. Now, how does this affect the glass machinery and inspection industry? The IS machines and inspection machines have to be very flexible to be able to handle any shape and size of container. To stay productive with short production cycles, job changes need to be done quickly no matter how different the containers are in size and shape. Inspection machines have to be able to adapt to a large ware range with few changes to the machine as possible. The changes that are needed to go from one container to another also have to be easy.

More importantly, the inspection machines also have to be flexible in the sense of expandability. For example, if a machine with the capability of performing up to five inspections is purchased, but at the time of delivery, only two inspections are needed, it should be possible to buy the machine with two inspections and add other inspections later. Also, as new technologies are developed, it should be possible to add them to an existing machine. A modern inspection machine should act as a ware handling platform with the capability of adding new inspections as necessary.

Ease of Use

The development of technology, as well as the growing demand on inspection quality and flexibility, makes inspection machinery more and more sophisticated. This growing technological sophistication helps to address shortages of skilled labor willing to work in the often-difficult environment of a glass plant. Therefore, it is important that, despite the increased complexity of inspection machines, the operation remains simple and intuitive. Furthermore, it's important that the machines operate reliably so troubleshooting can be kept to a minimum. Making use of today's commonly available local area networks, remote access to the machine from any location can assist in many aspects of troubleshooting. To support the ease of operation, especially at job change, several measures can be taken. As many job parameters as possible should be able to be stored and recalled. This includes positions and parameters of motors, camera setups, (such as focus, zoom, gain, contrast, and position), and inspection zone setups.

In short, everything should be automated to make job change and machine operation easy and operator independent. For repeatable inspections, it is particularly important for inspection setup to be operator independent. This means that no matter who sets the machine up, the inspection result is the same. If all parameters are stored in job files, it is possible to transfer the data from one machine to another or an expert can easily check setups. To be able to store setup parameters, it is necessary that as many movements as possible are motorized. Using servo and stepper motor technology makes this possible for nearly all motions.

Vision Technology

In the early days of automated glass making (and in some places even today), inspection was done visually by plant personnel. This human inspection has been partially replaced by mechanical inspection. Both inspection methods have their advantages. Human inspection is non-contact, flexible, adaptable, and delivers detailed information on the inspection. Mechanical inspection has speed advantages, as well as operator independence, higher reliability, and repeatability. The goal is to merge these two inspection methods into one superior inspection. The answer is machine vision inspection. It is one of the key components for a flexible, reliable and repeatable inspection line. It can replace the mechanical inspection while being faster and providing more detailed inspection information.

There are three components that form the cornerstones of vision inspection. These are cameras, optics/lighting, and image processing algorithms. Over the past few years, the rapid development of cameras and image processing systems offer new inspection capabilities never believed possible. Camera resolution has increased, allowing detection of smaller and smaller defects. Together, with an increase in image processing power, these new cameras allow for improved inspection quality while maintaining the necessary speed.

To be able to capture high quality images you need more than good cameras. You also need high quality light from the right direction, with proper brightness and timing.

Various arrangements of LED lights, mirrors, polarizers and beam splitters help get the right light to the right place. Controlling the brightness of the light by a computer offers the ability to perfectly adjust the light to different glass colors and thickness. Another very powerful lighting feature is the patterned lighting. By switching off rows of LED lights, defects that can't be seen with standard lighting now show up with surprising clarity (Figure 1).

Different arrangements of LED lights also can be used to make different defects visible in the same area of the container (Figure 2).

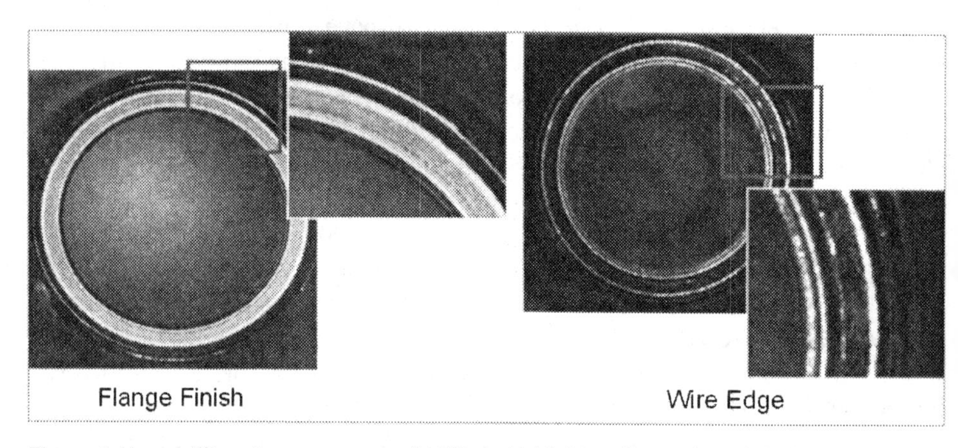

Figure 1. Use of different arrangements of LEDs to highlight sealing surface defects.

Once the high quality image is captured, algorithms are needed to find the defects on the image. Several different algorithms have been developed – and several more will be developed – to find all possible defects in any location of the container. For example, it is possible to find stones in the stippled area of the base when filtering out the regular pattern and looking only for irregularities.

High speed, high-resolution cameras, LED lighting, and fast image processing are important for a reliable vision inspection, but these elements aren't enough to ensure an easy setup and operation. To assist in setup, all camera and light adjustments need to be motorized and computer controlled. When changing from one container to another, it is necessary to adjust the position of the cameras when doing base or finish inspection. These adjustments can be time consuming if done manually. Motorizing the up and down movements of the cameras, as well as the lens adjustments and zoom settings, allows all of these positions to be stored in job memory for quick recall when changing back to a repeat job.

For sidewall inspections, the use of high-resolution cameras eliminates the need to make adjustments to the camera settings. For different container sizes, it is possible to simply use a bigger or smaller section of the whole image.

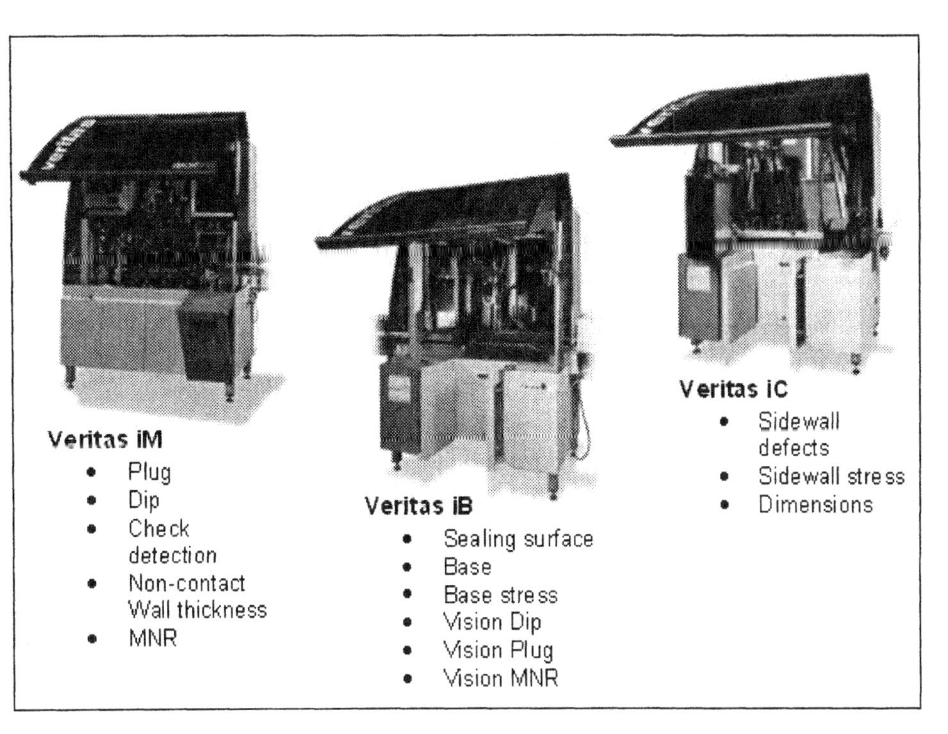

Figure 2. Emhart Glass Veritas series inspection machines.

Motorizing all adjustments offers additional benefits other than making setup faster and easier. It also helps make inspection more repeatable and reliable. Once a job has been set up and stored by an expert, it can be recalled easily. Job set up becomes operator independent.

As mentioned above, machine vision technology has gone through a rapid development, which still is ongoing. Inspections that were previously done mechanically now can be done visually, and there are more to come. Two such inspections are plug and dip inspection. When done mechanically, both inspections have the potential to damage the container through physical contact. The visual inspection does not touch the container. It gives more detailed defect information with higher precision.

Automation

Advanced inspection technology, combined with advanced software packages, bring inspection to a level that goes beyond just finding defects. An increased demand for productivity and difficulty finding skilled labor has led to a demand for automation in the glass making process. The goal is hands-off, lights-out production. Inspection machines can be an integral tool in reaching that goal. Several enhanced features can be incorporated into the inspection machine, thus helping operators do their daily work. One feature already mentioned is remote diagnostics and maintenance. This feature makes use of the fact that, today, almost every computer is linked to a global network or can be connected at any time through a modem. This gives access to the inspection machine from any location in the world. Using up-to-date security tools, the inspection machine can be protected from viruses and unauthorized access. This type of remote access enables experts to check the status of the machine, make changes to the setup, assist in operator troubleshooting and update the software.

Another feature worth mentioning is the need for inspection reports. Most defects found in the cold end have a known cause. If properly reported, defects could be eliminated which would lead to a higher pack rate and productivity. Inspection machines can correlate all inspections to a mold number and can report detailed information on the findings to the appropriate location via the plant's internal network. For example, findings can be reported to the hot end, the source of most of the defects, or to the quality department for technicians to take action. Another possibility is to collect the data for statistical purposes. Provided that the hot end has an intelligent control system, corrective actions can be taken on the information from the cold end or maintenance personnel can be alerted to repair a defective part.

Another feature that needs to be added is a semi-automated or fully automated sampling mode at the cold end. Several times a day, every inspection machine needs to be checked to make sure that it reliably detects all defects. If not, the inspection has to be re-calibrated. Typically, this machine check is performed manually. An operator takes a set of sample containers with the defects, runs them through the machine and checks if they were rejected for the correct reason. The inspection machine can assist with this procedure by stopping the ware upstream of the machine and allowing sample containers to pass through the machine. During the inspection, the machine provides information for rejecting defective containers and stops the containers at the end of the machine. Going a step further, it would be possible to place the containers on a sample loop and, whenever necessary, the machine will stop the ware flow and inspect the sample containers from the sample loop. Provided the machine contains the necessary information on the sequence of every defect, the machine can perform a self-check and alarm the operator if something is outside the norm.

Veritas

Emhart Glass took the challenge of the industry and created a new inspection family. We took state-of-the-art technology and built three inspection machines that cover the full range of inspection. They are flexible, reliable and repeatable.

Veritas iM

The Veritas iM is a starwheel handler that performs mechanical plug, ring and dip gauging. In three rotating stations it does check detection, non-contact wall thickness and heel dot code mold number reading. The plug, dip, starwheel and infeed screw motions are all very dynamic and require high precision and fast movement. These are motions that are ideally done by servo motors. The three rotators carry out very simple movements and can be driven by inverter motors.

Veritas iB

The Veritas iB is one of the two vision inspection machines modules of the Veritas family. It performs base and base stress inspection, sealing surface inspection, and vision plug and dip inspection. The Veritas iB also has a mold number reader that correlates all inspection to a mold number.

All adjustments on the Veritas iB are motorized. This includes belt handler and camera height adjustments, as well as camera lens adjustments. This makes it possible to job change without opening the cover of the machine with the touch of a button.

The Veritas iB also makes innovative use of vision technology to perform vision plug and vision dip inspection. This vision module can replace the mechanical inspection by a non-contact version, which eliminates the danger of damaging the container when inspecting it. Furthermore, this allows inspection speeds of up to 600 bpm.

Veritas iC

The Veritas iC is the third module of the Veritas family. A fully non-contact vision inspection machine, the Veritas iC performs all sidewall inspections. This includes detection of opaque and transparent sidewall defects, sidewall stress, and dimensional inspection. The dimensional inspection consists of body diameter, lean, and height inspection. Six cameras are used to enable the vision system to get a full 360° view of the container. As mentioned earlier, the lighting is very important to get good inspection results. One of the unique features of the Veritas iC is its patterned lighting. This lighting allows finding defects not visible with normal lighting. The Veritas iC light boards consist of several thousand LEDs which can be computer controlled to adjust the brightness to the color of the container. Every camera takes two images of the container providing a total of 12 images per inspection. Between the two image acquisitions, the lighting changes in one inspection station and is polarizing the light in the other inspection station.

Conclusion

Emhart Glass analyzed the glass market and, based on this information, created a new inspection family that utilizes the latest technology. As a result, Emhart Glass has developed a product line that is flexible, reliable, and repeatable. More importantly, it is positioned for the future.

Quality and Glass Production Improvements Through Statistical Process Control at Fevisa in Mexico

Jesus A. Ponce de Leon, JPdeL and Associates (USA) and
Juan Rafael Silva-Garcia, Fevisa Industrial (Mexico)

Abstract

This paper presents the efforts and lessons being learned in applying the techniques of Statistical Process Control in a production line from furnace to packing. The objectives of these efforts are aligned with the strategic priorities of the company. These objectives are learning, quality improvement, better efficiency and pack to melt ratios. The paper also explains how these techniques are being used during the implementation path to bring the organization from an art and craft environment to a professional world class level of performance.

Introduction

The quality movement started in Japan by Deming in the early 50s, spilled over the US during the early 70s. The Western Electric company, is perhaps among the very first American companies to heed Deming's call for statistical control and management leadership. Their Statistical Control Handbook, sparked the interest of others such as AT&T and aerospace companies such as Boeing and Lockheed to improve quality through SPC. Since then we experienced an evolving slew of tools and techniques from the simple problem solving tools, to quality circles, to statistical control, to six sigma and some others in between. Many organizations have reported significant improvements in their quality and the bottom line due to the efforts and application of these tools and management philosophy. Testimonials of the benefits of a quality improvement orientation have been presented in conferences and compiled in conference memories from ASQC, AQP, the Baldrige Award, and the many state quality awards and manufacturing associations such as the Illinois manufacturing Association Productivity and Quality Conference (see, for example, Profiles in Quality, 1991, Kinni, 1996).

The management of Fevisa Industrial SA de CV, has understood the importance of adopting a customer and quality orientation and began a more integrated organization-wide program in 2002. This paper presents only a small portion of this application of SPC to improve process control, to improve operations, and decrease defective product.

Foundations

It is fashionable nowadays to be involved in a six-sigma program even if the one adopted is not exactly a full fledged six-sigma program. SPC is an integral part of a six-sigma program and one of the many tools utilized to produce significant and measurable customer driven requirements and reduce quality defects (see for example Breyfogle II, et al 1999, 2001).

Dr. W. Edwards Deming found a receptive student of quality improvements in the Japanese in the late 50s. His approach to quality fundamentally was based on management intervention to create constancy of purpose toward improvement, to build quality into the product (prevention), and in identifying systemic and special causes to reduce process variability and consequently improve productivity and quality (for a detailed list of Deming's 14 points and tools see Breyfogle II, et al, 2001, Garvin, 1986, Walton, 1990).

Dr. Deming recognized that all processes vary thus statistical control tools were needed to understand how they vary and to fully distinguish systemic from special causes (Walton, 1990). He advocated management responsibility to help people to participate in the decision making and to make them smarter in solving problems and learning (Walton, 1990). Arguably these are also key elements of a six-sigma program.

At Fevisa Industrial, an integral program has been launched named New Working Culture (Nueva Cultura de Trabajo in Spanish) that incorporates learning, statistical tools, 5s, quality and problem solving tools (seven quality tools), decision making training, supervision training, information technology, and reorganization and enhancing of the quality, laboratory and auditing functions. The program has been launched not as a fad expecting quick results and then move on to another flavor of the day project but as a sustainable organization-wide effort that involves all the employees and management levels. Figure 1 shows a simplified version of this program.

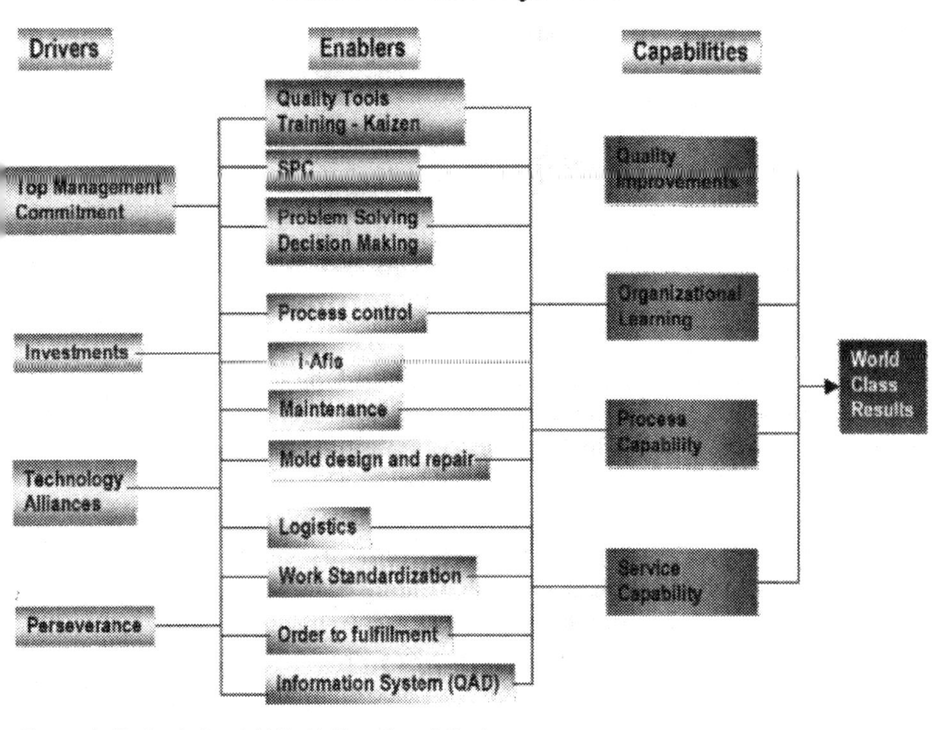

Figure 1. Fevisa Industrial World Class Level Goal

Top management has set a stretch goal to reach world class performance levels within seven to ten years. The capabilities to be developed to reach and sustain this goal are: quality and process improvements, organizational learning, process capability and service capability. These in turn are supported by the enablers listed in the middle column of the diagram SPC being one of them.

SPC a Key Enabler for Change

Fevisa Industrial is a niche player in the glass industry that has a well focused strategy. It has chosen to serve a few strategic customers rather well. Due to its size it can not play the economies of scale game. It has to excel in its operations and quality of its products and service levels. In order to do it Fevisa Industrial has invested in more productive technologies such as triple gob, bulk packaging, information technologies to improve decision making and have real time operations performance information and more recently on process control technologies.

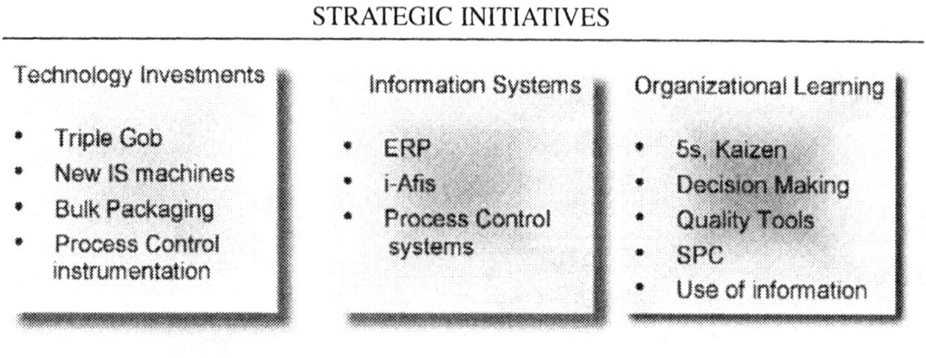

STRATEGIC INITIATIVES

Technology Investments

- Triple Gob
- New IS machines
- Bulk Packaging
- Process Control instrumentation

Information Systems

- ERP
- i-Afis
- Process Control systems

Organizational Learning

- 5s, Kaizen
- Decision Making
- Quality Tools
- SPC
- Use of information

Equally, if not more important, it has invested in a long term program of organizational abilities focused on its employees and their learning. This later track is comprised, as shown in figure 2, of various tools and methodologies SPC being one of them.

In the last three years Fevisa Industrial began to generate data in a more systematic way from its production processes (furnaces, forehearts, IS machines, lehr, cold end). In the first stage of implementing SPC a customer requirements table was developed and from these customer requirements FI proceeded to elaborate a detailed process map identifying the Key Process Input Variables (KPIV) and Key Process Output Variables (KOPV). Once these had been identified a systematic collection process, step two, was put in place and after a year of data collection and of keeping track of major variables the second stage, data analysis, began.

With the process by process KPIVs and KPOVs, defined in step 2 , a systematic comparison was made with the real and actual process variables that were recorded on a periodic basis. Steps were taken then to begin recording the KPIVs and KPOVs not being systematically recorded.

Step four, was to study and analyze the run charts, ask experts in the field, benchmark against technology partners and review of original machine and process documentation to identify the first run charts upper and lower limits. Of course, customer requirements were also a key input for this task.

Statistical Process Control Track

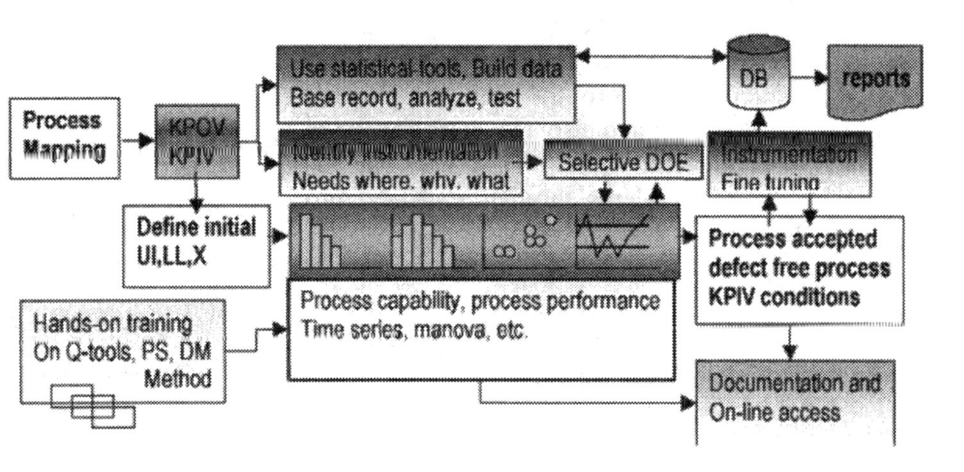

Figure 2. Elements of the statistical process control program

To prepare key employees on reading data FI organized training sessions on 8D problem solving methodology and in the seven quality tools. Leading teams were formed to apply these tools to finding root causes and reducing quality defects. There was a slowdown in this component of the program after a year of active formation of teams due to a couple of factors, the installation of new technologies, for example replacing old IS machines for newer ones, and the turnover of some key employees. Team problem solving and quality improvements resurfaced again this year.

To support the SPC effort and learning a computerized information system (i-Afis) has been implemented recently to collect production and defects data in real time from the IS machine to the cold end (Inex, TIM, etc) building a data base that can be consulted any time. Previously, product quality defects data were available in a closed circuit display at real time but not kept in memory unless someone will take a written note of the data. Still, key variables (process variables such as temperature in the furnace, foreheart, air pressure, etc.) have to be recorder on a periodic basis in the shift logs.

FI is at this moment training its personnel in the principles of variability, elaborating run charts, histograms, while the quality function department prepares daily and then weekly the SPC charts and trends as they relate to key variables and main defects per machine, section and cavity. Taken the organization this far has required perseverance, strong top management support, and little by little incorporating these concepts and visual data into the daily production meetings. A special quality trends meeting was added solely dedicated to analyze quality defects per machine.

Other glass companies have already gained some experience with SPC. Corning Glass Works (New York) began using SPC as far back as 1990 as a mean to providing the tools for the operators to determine when the process was out of statistical control and also as a mean to take responsibility for corrective actions. A third objective was to involve their operators as quality inspectors by preventing quality problems through statistical process control (Profiles in Quality, 1991; Kinni, 1996, 202-203).

FI has adopted Minitab as the software for data visualization and statistical analysis. The software is primarily used by the quality department. As part of the training program it will taught to operation teams.

The transfer of SPC knowledge has been slow due to the technical nature of SPC and the inexperience of the line personnel with statistics. Thus, the training has been spread into several stages. In the first stage external experts in conjunction with the quality department engineers have been collecting, analyzing and running the process control data, one application is presented in this paper. In the second stage, training through a gradual immersion in the concepts of statistical process control is underway. A third stage requires learning how to read the charts, basic statistics and patterns of data points in the run charts. A fourth stage will engage line personnel in preparing themselves the charts and interpret them. In the final stage the operators should practice how to read the charts and take responsibility for maintaining under control the KPIVs and observe their effect on the KPOVs. It is estimated that still between one and two years will be necessary to reach these goals.

An Application of SPC in Fevisa Industrial

The following is an excerpt of how SPC is being applied to furnace three, its machines and packing lines and being rapidly extended to all other furnaces and their respective machines and cold ends.

The overall process of analysis is divided into three major steps:

- Graphical analysis of the behavior over time of process control variables from furnace all the way to IS machines.
- Statistical analysis of the process variables from number one above.
- Analysis of the technical differences between lines under analysis.

The first three graphs show how systematically external noise was removed to identify the natural (inherent) performance (behavior) of the pack to melt line. A graphical format like this is easier for line personnel to understand what a special source of process variability is, and what a natural process variance, clean of major external noise, is like.

The result of the variance caused by external sourcing of electrical power can be seen by comparing graph 1 vs graph 2. Both graphs plot pack to melt efficiency. Removing the major external shocks leaves the natural capability of the equipment. The difference between graphs two and three are the major actions that the personnel take on a daily basis in attempting to correct quality problems, the effect of equipment failures and corrective maintenance to parts, equipments and support systems such as compressors. Once these effects are removed the process capability behavior is revealed (graph 3). This systematic and slow process of cleaning the data from major external noise is necessary in order to run the process variables against the quality defect levels (defects are directly correlated with pack-to-melt efficiency).

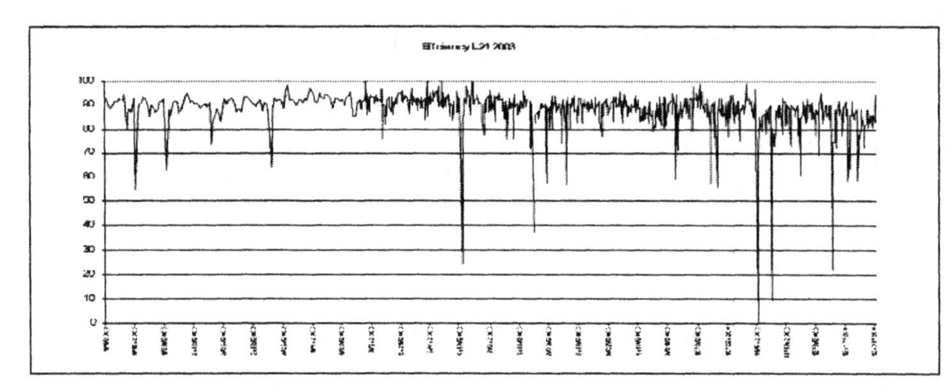

Graph 1. Efficiency L21 2003.

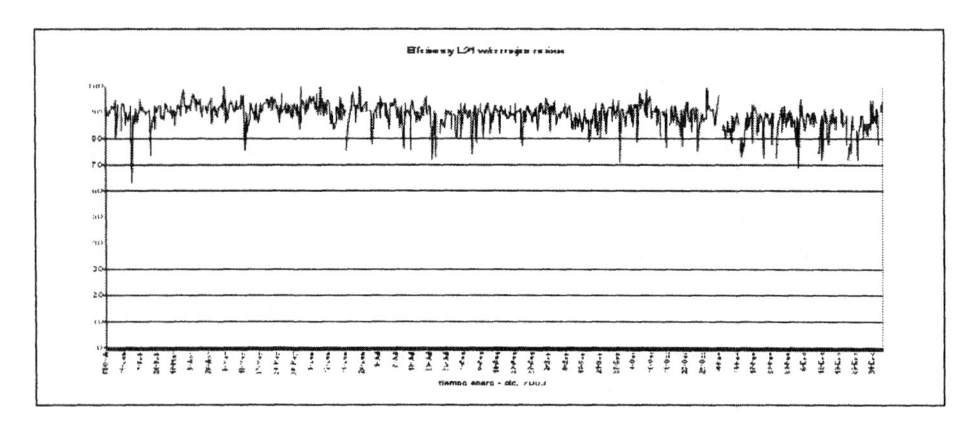

Graph 2. Efficiency without major noise.

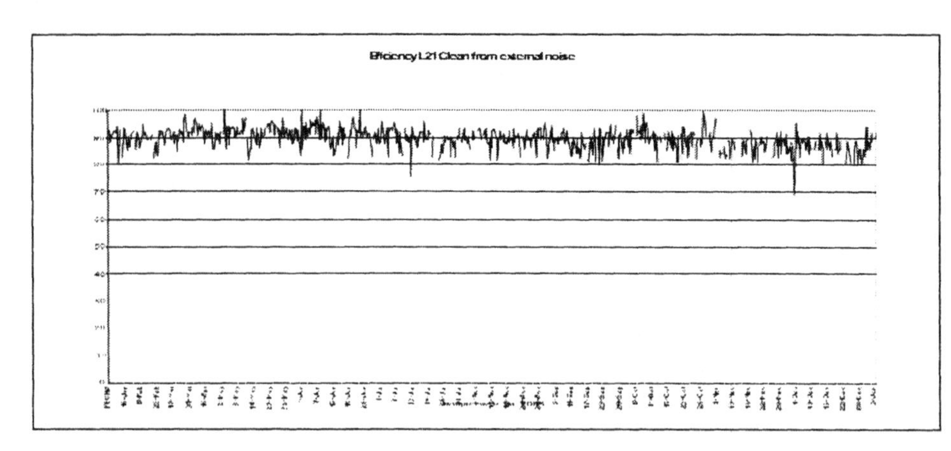

Graph 3. Efficiency L21 clean from external noise.

The following table shows the valuable information that is drawn from the graphical analysis.

TABLE 1. VITAL STATISTICS OF PACK-TO-MELT EFFICIENCY AND VARIANCE LIMITS.

Líne	Real Efficiency 2003 A	Efficiency without major Electricity plant-wide events B	Efficiency without Minor operational and Maintenance events C	Observations
L22	Average 88.2% 1 sigma 6.52% Limits: Upper 88.3+6.5= 94.8 Lower 88.2- 6.5= 81.7	Average 89.2% 1 sigma 3.36% Limits: Upper 89.2+3.4= 92.6 Lower 89.2-3.4= 85.8	Average 89.7% 1 sigma 2.46% Limits: Upper 89.7+2.4= 92.1 Lower 89.7- 2.4= 87.3	B – A = 1 % effiiciency C – B = 0.5% efficiency
L21	Average 86.9% 1 sigma 3.19% Limits: Upper 86.9+3.2=90.2 Lower 86.9 -3.2=83.5	Average 87.5% 1 sigma 2.07% Limits: Upper 87.5+ 2.1= 89.6 Lower 87.5- 2.1= 85.4	Average 88.6% 1 sigma 1.67% Limits: Upper 88.6+1.7= 90.3 Lower 88.6 -1.7= 86.9	B – A = 0.6 % efficiency C – A = 1.1 %
	Plant wide electricity shutdowns affect L22 more than L21 causing failures in blowers, servors, motors and electronic circuits.	L21 recovers faster from major failures than L2. Moulds, blanks and mechanisms tend to be more affected in L22 than in L21.	Our internal benchmark is L21 due to being fed by same furnace and due to its better performance. Important differences start in the foreheart.	Some differences: Temperature profiles in foreheart zones and orifice (feeder). Technical modifications in the orifice area. Technical changes made to moulds and blanks. Changes made to the electronics.

The second step is to use these results to better understand the potential gains and behavior of the equipment after an external shock has affected its production. Line employees can also see how much variance really exists in the process once the major external noise has been removed. Several goals emerge from this analysis: levels of high efficiency that can be reached, how ample is the variance of the equipment, how well it behaves when under control, and how long it takes to recover from a major external shock (it was found it takes three days to restore its natural process capability).

In the second step, a statistical analysis is performed with Minitab running a series of regressions, scatter data and goodness of fit against data normality. Once the data have been cleaned of major noise and outlier data points statistical tools are used to find major effects of KPIVs on pack-to-melt variability and performance levels. The following table shows the results of such analysis.

A number of regressions were run testing the quality of the data (find co-variances and eliminating them), eliminating outliers, eliminating incomplete data rows, and testing for normality of the data. The variables tested were:

- Furnace temperature zone 1
- Furnace temperature zone 2
- Throat temperature
- Channel entrance temperature (one for each arm)
- Foreheart zone 1
- Foreheart zone 2
- Foreheart conditioning zone
- Orifice temperature
- Gobs temperatures (inside, middle, outside)
- Blank temperature
- Finish mold temperature
- Cooling Air (pressure and vacuum)
- Plate temperature
- Plunger height

These variables (KPIVs) were regressed against total defects per line (pack-to-melt) and against major defects (leaners, line overs, and others). The results are shown in the table above (simplified for this presentation).

The analysis provides insightful information. First, the variance explained by the most statistically significant KPIVs revealed that only about 14% of the total variance in defective is explained by the process variables. That leaves a large portion to other causes. After an analysis of what factors have been left out of the analysis it was concluded that setup quality, mold shop repair, working condition of fixed and variable mechanisms and machine timing setup were most likely to be these additional sources of defective. Thus, it was concluded that a systematic analysis of the differences between IS machines, between fixed and variable mechanisms, machine timing setup, and mold repair and quality of setups needed to be done and compare to our technology partner recommendations and benchmark against other operations.

This analysis was prepared by developing a matrix containing each mechanisms and each major detail of the IS machine such as space between molds, position of vertiflows, position and angle of channels etc. This systematic comparison still is underway. In the process a number of important differences have been found that might explain why mechanisms may be wearing out faster, why inverter arms are vibrating, why molds are getting nicks and dents, and many more. This analysis also pointed to the mold shop as a major potential source of the quality problems.

FEVISA REGRESSIONS RESULTS FOR 2003
July 1 to December 16, 2003

Variable	L11 Average	L11 Variance	L12 Average	L12 Variance	L21 Average	L21 Variance	L22 Average	L22 Variance
%Defective	**11.48**	**3.21**	**11.91**	**3.51**	**12.88**	**3.19**	**20.04**	**4.99**
FurnaceZ1	1235	22.38			1238	22.9		
FurnaceZ2	1207	12.41			1258	23.5		
FurnaceZ3	1074	11.07			1244	17.6		
Throat (*)	1046	4.84			1089	12		
Channel entrance	1050	4.82			1010	8.17		
ForeheartZ1	1208	1.57	1222	45.31	1217	8.6	1198	4.57
ForeheartZ2	1210	0.94	1218	7.71	1205	2.9	1199.8	3.69
ForeheartZ3	1199	4.13	1221	60.91	1187	11.26	1196	2.59
Conditioning Zone			1148	5.7	1190	7.12	1117	29.15
Orifice Ring					1129	1.78	1186	6.26
Gob Inside	1178	4.28	1176	3.52	1175	3.9	1171	3.32
Gob Middle							1185	4.65
Gob Outside	1183	5.64	1180	4.31	1179	5.1	1183	6.41
Speed bpm	295	4.1			293	2.9		
Air Temperature								
Final Blow					42.5	3.24	40	3.48
Vacuum Press					15.4	1.6	18.9	2.26
Temp Blanks								
Plunger								

		L11		L12		L21		L22
Variables that		FORE Z2		GTI		ORIF RING		ORIF RING
Explain the		FORE Z1		FORE Z1		COND ZON		FORE Z2
regression		GTO				FINAL BLOW		FINAL BLOW
R-Square		**14%**		**12.50%**		**24.60%**		**12.10%**

Total variance explained by the selected variables above

One-Tail	L11 Beta	L11 P-Value	L12 Beta	L12 P-Value	L21 Beta	L21 P-Value	L22 Beta	L22 P-Value
Datos	**N=108**		**N=109**		**N=110**		**N=71**	
Variable								
FurnaceZ1	0.067				0.225			
FurnaceZ2	0.006				0.218			
FurnaceZ3	-0.201	0.076			0.191			
Throat	-0.097	0.072						
Channel entrance	-0.106	0.066	0.186	0.026	0.221			
ForeheartZ1	-0.093	0.02	0.168	0.04	0.097	0.145	0.18	0.067
ForeheartZ2	-0.111	0.001	-0.03	0.378	0.033	0.358	0.214	0.037
ForeheartZ3	0.05	0.16	-0.058	0.275	-0.019	0.228	0.013	0.456
Conditioning Zone					-0.289	0.01	0.062	0.241
Orifice Ring					0.353	0.005	0.195	0.052
Gob Inside	0.125	0.195	-0.076	0.215	-0.002	0.49	0.012	
Gob Middle							-0.008	
Gob Outside	0.005	0.011	0.073	0.226	-0.237	0.043	-0.07	
Air Temperature							-0.161	0.091
Final Blow					-0.136	0.16	-0.276	0.01
Vacuum Press					-0.303	0.014		
Speed bpm	295	0.001						
Temp Blanks								
Plunger								

R-Square = explains how much of the total variance is due to the statistically significant variables.

P-Value = statistical significance of the variable. If this value is less or equal to 0.05 it is entered into the equation as a significant contributor to the expected result of the dependent variable (quality defects)

Beta = Slope and direction of the relationship between the variable and quality defects.

The regressions resulted in the following variables being significant in explaining quality defects:

L22	(1) Orifice Temperature	(at higher temperature more defects created)
	(2) Temperature Foreheart Z2	(operating at higher temperature causing defects)
	(3) Pressure of final blow	(insufficient air to cool down causing defects)
L21	(1) Orifice temperature	(at higher temperature more defects created)
	(2) Conditioning Zone temperature	(cold temperatures related to quality defects)
	(3) Pressure of final blow	(insufficient air to cool down causing defects)

A series of actions were launched to document work procedures, mold repair inspections, work flow, data recording, quality of operator's work, measuring instrumentation and their calibration, and review of blueprints of molds and blanks. It was found that a number of important deviations in work procedures, setup quality and repair quality were in need of standardization. A program was launched to standardize work procedures for repair and reception of molds and blanks. A data base has been created to track the type of physical damage that molds and blanks show after being removed from the machine and what problem were they been possibly causing. Training and a recalibration for all measuring instruments in the shop program was also launched.

A sample of other conclusions that emerged from the three pronged statistical analysis is presented below:

1. From the analysis of the graphs on efficiency it is concluded that actions (possibly investing in equipment) to minimize, or dampen the effects of plant wide electrical failures is of high importance.
2. The opportunity from action one is to increase x% of efficiency in L22 over a year.
3. We see that the aftershocks from plant wide electrical failure cause similar chaos causing failures in servos, blowers, electronic circuits, damaged moulds, blanks and mechanisms due to insufficient compressed air to complete three more cycles.
4. The second group of actions should be directed at verifying setups, moulds and blanks being within the right specifications and installed properly.
5. Maintenance is going to review the data on motors, servos, circuit boards, critical instrumentation and so forth to identify the pareto of failures and their likely recurrent causes in the after shocks.
6. In terms of process control, the fabrication department is going to homogenize the glass to reduce first the ample range between the gob temperatures and their variance.

7. Once this is done then fabrication will proceed to reduce the temperature of the gobs keeping track of the operating conditions of the other variables such air pressure, etc.
8. A detailed analysis of the technical and design of the blanks, moulds and mechanisms will be continued to identify constraints in cooling and temperature profiles in the forming of the bottle.
9. A visual board will be posted in the areas that will integrate the information of key control variables and their trends highlighting the limits of controls where the variable should be operating.

10 Temperature control within narrower ranges need to be established in furnace 3 especially in the foreheart in all three zones and the conditioning zone temperature in particular, the question is how wide should the limits be and what level shall be selected.

Pending for Further Analysis

1. To include other variables, especially IS related (mould conditions, blanks conditions, setups, wear and tear of mechanisms).
2. Regress gob position (inside, center or outside and machine position) with defects and type of defect.
3. Correlate blanks and mould and mechanisms changes to defect level changes by position and over time.
4. Run ANOVAs to identify key interactions between relevant variables (crossed effects between any two variables especially between those related to the channel and those related to the IS machine). This is the Design of Experiment (DOE), by taking advantage of the data that we have already collected and is being updated to 2004.
5. Furnace's temperatures need to get under control to reduce its effect on the subsequent processes (foreheart). See attached temperature profiles for 2003 below.
6. Continue with the analysis of technical and design differences between L22 and L21 establishing hypotheses of likely cause-effect.
7. Continue with the in-depth review and redesign of the information and process flow in the mould shop, metrics, follow up systems and quality of work done.
8. Identify some of the best performance benchmarks internally and with technology partners to establishing the bands for desired control of key variables.

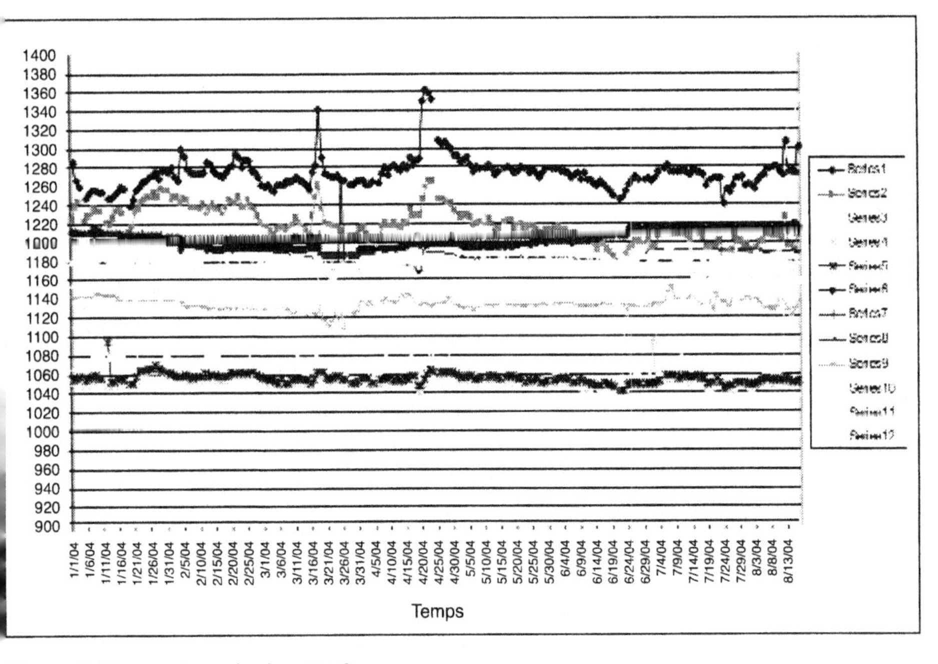

Figure 3. Temperatures for furnace 2.

Conclusions

Fevisa Industrial recognized the value and importance of learning and decision making with facts. Although SPC and quality initiatives have been around in the USA for many years, and in Mexico since the mid 80s, it is still a nascent methodology in many manufacturing facilities. The glass industry has been characterized as art. It is becoming more evident with the passing of time that new technologies require of fine tuning of process variables and of narrower faster controls to eliminate wide temperature variance for example.

The experienced gained in implementing a program based on quality improvements and process control based on facts rather than pure experience shows that an integrated program of technological initiatives, quality tools, statistical analysis and learning are critical to achieving high levels of performance. These initiatives require time and consistent support from top management. The pressure to drop these long term programs is strong when on a short term basis operations may not seem to sustain their improvements. But over the long run is becoming evident that accumulated learning, fact based decision making and SPC are necessary to understand why things happen, how they happen, and what can be done to correct it.

People are always the major factor in the implementation of such programs. SPC is a difficult tool for many, especially if their education level is not a four year engineering degree or they have been out of school for a long time. Learning by doing is important but learning by understanding, in particular understanding why process control is critical, is even more valuable. Learning by experience will require many years of wait expecting that our employees have an opportunity to see all sorts of problems hoping that they will remember how it was solved, and what was done. But memory alone is not reliable in the long run. Data have to flow constantly from the floor, be recorded to be able to analyze it and play with it, to test it, and develop hypotheses of cause and effect. Simple quality tools are also important to prepare and arm our employees with the discipline for problem solving, data collection, and more importantly for data analysis. The emphasis is on learning the whys, the cause and effect relationships.

SPC is in its early stages of implementation and has taken some time for employees to understand it, what it does and why it is useful, most important is that they understand that a process out of statistical control is a process that will cause quality problems. Visualization of key variables (KPIV) and their effect on quality (KPOV) are starting to sink in their way of working. Once they have understood how to read and what it means to be in or out of control they will proceed to take immediate action to correct or prevent key variables to be out of control. This mentality is beginning to permeate not only the hot end (fabrication) but also the mold shop and the cold end operation.

The major roadblock during the implementation of such efforts has been the battle for time. Key employees are constantly distracted away from the systematic application of these techniques for the daily special problems. Deming called our attention long time ago about this struggle. Management is responsible for removing systematic problems of quality while workers should take responsibility for solving and maintaining a process under control. Perseverance is a virtue in this effort, one that top management can not abandon if techniques such as SPC are to be a part of our employees' work habits and source of better decision making.

Update 2005

By late 2004, SPC practice and understanding of what process control means on the floor have begun to produce consistent results in all machines. The following table summarizes the vital statistics of four lines (some numbers have been modified for confidentiality but the essence has been retained). Not only the mean performance of each machine has significantly improved but also its variance has been greatly controlled.

Efficiency	2004			2005			t-Test	
	N	AVG	STDEV	N	AVG	STDEV	T	p
L1	70	84.02	4.92	69	90.29	3.09	-8.98	0.0
L2	74	81.07	10.49	67	88.07	2.58	-5.32	0.0
L3	77	84.23	5.57	58	88.95	3.21	-5.77	0.0

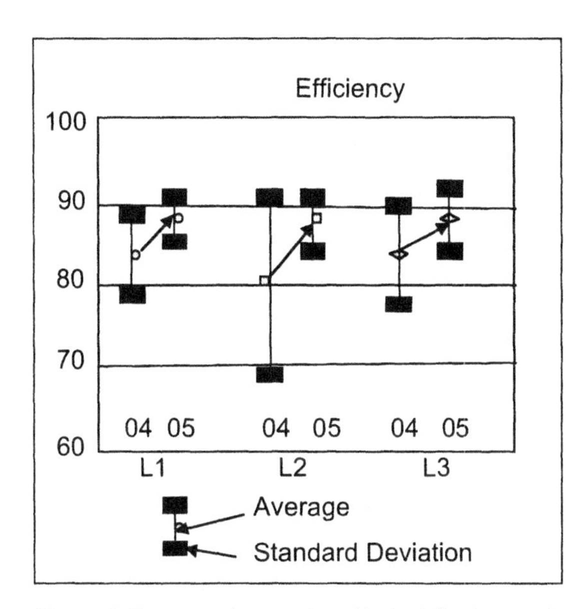

Figure 4. These results are plotted in the following graph visually showing the improvements made through SPC.

BIBLIOGRAPHY

Breyfogle, II, Forrest W. Implementing Six Sigma: Smarter Solutions Using Statistical Methods. John Wiley & Sons. New York. 1999.

Breyfogle, II, Forrest W. J.M. Cupello, Becki Meadows. Managing Six Sigma: A Practical Guide to Understanding, Assessing, and Implementing the Strategy that Yields Bottom-Line Success. John Wiley & Sons, Inc. New York. 2001.

Garvin, David A. A Note on Quality: The Views of Deming, Juran, and Crosby. Harvard Business School case 687-011, 1986.

Kinni, Theodore B. America's Best. John Wiley and Sons, Inc. 1996.

Profiles in Quality: Blueprints for Action from 50 Leading Companies. Bureau of Business Practices, Allyn and Bacon, Mass, 1991

Walton, M. Deming Management at Work. G. P. Putnam's Sons. New York. 1990.

Application of Microwaves in Glass Conditioning

Peter Vilk, BH-F (Engineering) ld, Didcot, Oxon, UK

Results from tests on several forehearths are presented. The effects of microwaves on chemical and thermal homogeneity can be observed. Glasses exposed to microwave energy include borosilicate, soda-lime, barium and lead crystal.

Introduction

The idea for application of microwaves in glass conditioning came from observing a 10-kg MW melting furnace in operation (see figure 1). The MW melting technology developed by Dr Hajek is an innovative non-contact heating method based on generation of high intensity MW field focused into a small space to start melting process locally from cold.

The MW melting is first, volumetric and selective method with an inverse temperature profile. Glass is heated from inside-out i.e. in opposite direction to all other known non-contact heating methods. The heat transfer comparison is shown in figure 2.

In order to be able to observe and to study possible effects of MW on molten glass in a forehearth BH-F have constructed a test rig, as illustrated in figure 3. Using standard forehearth gas burners, the rig was brought up to temperature and green bottle cullet has been melted.

The rig was 915mm (3′) long, 406mm (16″) wide internally, with a glass depth of 152mm (6″). The gas burner manifolds were removed and suitable shielding was installed. A total of six MW generators were switched on giving a total power of 9 kW. It was observed immediately that microwaves are absorbed by the molten glass equally through the depth. Based on this encouraging observation it was decided to proceed with MW tests on several forehearths under normal production conditions. Tests were carried out, under normal operating conditions, on barium glass, soda lime glass, lead crystal and borosilicate glass.

BARIUM GLASS TEST

Three microwave generators with a total power of 9 kW were installed through the roof of an electrically heated forehearth~~in~~ place of stirrer mechanisms (see Figure 4). Tri-level thermocouple readings were recorded. The forehearth pull was 18 tpd. It was observed that MW energy penetrates to bottom glass quickly with 40% of the installed power being absorbed. Bottom glass temperature responded to power changes almost as quickly as the surface glass (see Figure 5). Cord appeared to have broken up and dispersed. This first test showed that microwaves could be successfully used to heat glass in a forehearth.

Soda–Lime Glass Test

This was again an installation on a forehearth with radiant heating elements, as shown in Figure 6. The forehearth pull was 12 tpd. The total MW power installed was 6 kW. This was a short test.

The wave-guides were too far away from the glass and only 19% of the installed power was absorbed by the glass. There was no noticeable change in glass quality related to cord.

Lead Crystal Test

A standard construction of electrically heated lead crystal forehearth with submerged roof, tin oxide electrodes and electrically heated muffle was used in this test. The forehearth pull was 7 tpd. Two microwave generators with a total power of 6 kW were installed replacing stirrers (see Figure 7). Too much energy was absorbed by zirconia containing refractories~~on~~ly 7% by glass. Tri-level thermocouple readings are shown in Figure 8. Another trial is planned.

Borosilicate Glass Test

This test was carried out on an electrically heated forehearth with molybdenum electrodes, electrically heated muffle and submerged roof tiles. The forehearth pull was 16 tpd. The magnetrons were installed through the bottom part of the forehearth before the control thermocouple and the bottom drain (see Figure 9). The wave-guides were in direct contact with the bottom of the channel from the total of 8 kW power input 36% was absorbed by the glass, 45% by the channel material (Zarock). The main purpose of this trial was to study microwave energy effect on Zconia containing cord. Gob samples and samples from the forehearth drain were taken at regular intervals and analysed in a laboratory. The results were very encouraging. There was an overall reduction in visible cord and a very significant fourfold reduction in the diameter of cord strands.

Conclusions and Future Developments

From the limited, but encouraging, results of the tests carried out on live forehearths, under normal operating conditions, it can be concluded that:

- Molten glass in the forehearth does absorb microwave energy equally through the depth.
- The size of cord, containing zirconia, can be significantly reduced.
- Future plans include the repeat of tests on soda-lime and lead crystal lines.
- Future Developments can be outlines as follows:
- Upsize and industrialise the equipment.
- Provide engineered solutions for permanent installations to improve chemical homogeneity (reduce cord and "cat scratch") and thermal homogeneity in a forehearth.
- Attempt frit melting in colorant forehearths.

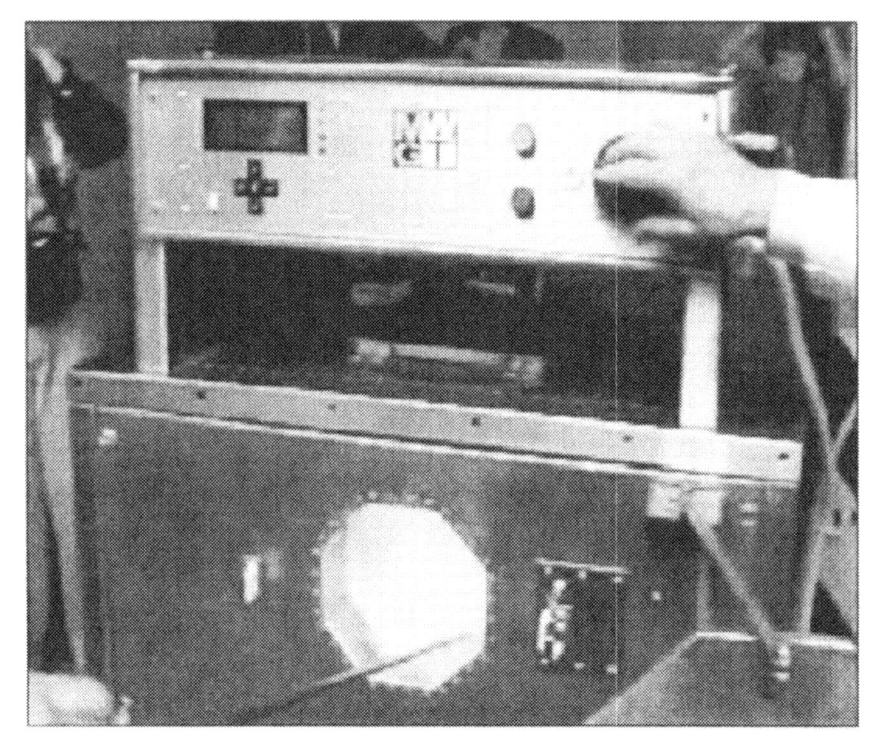

Figure 1. 10kg melter.

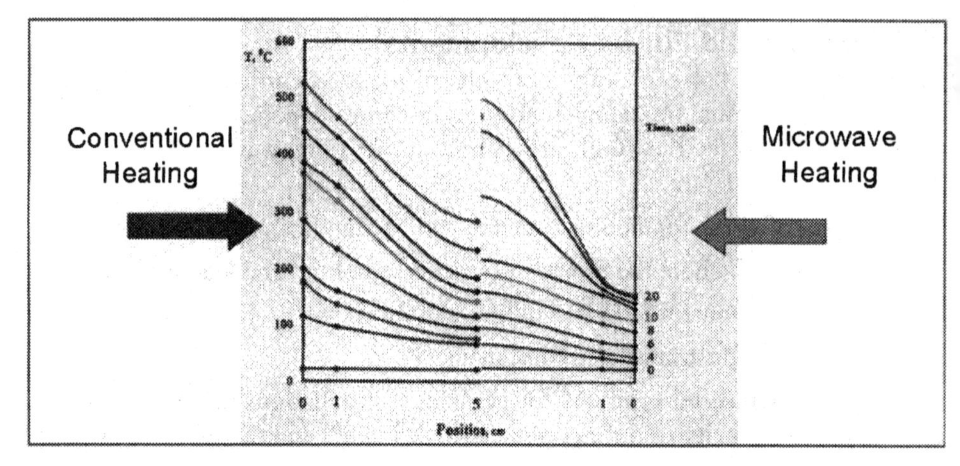

Figure 2. Heat transfer.

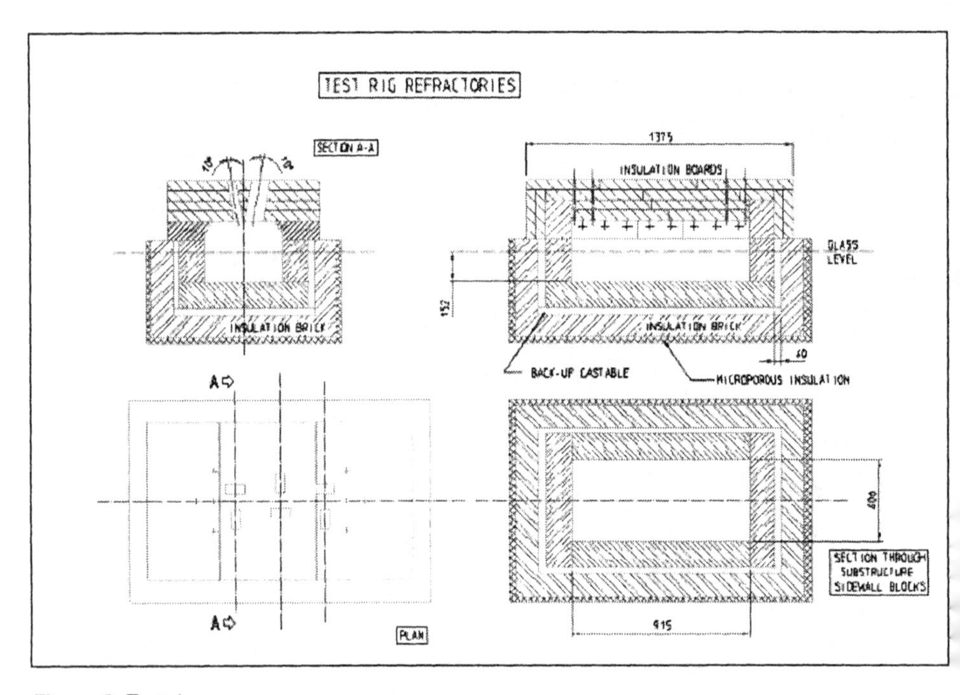

Figure 3. Test rig.

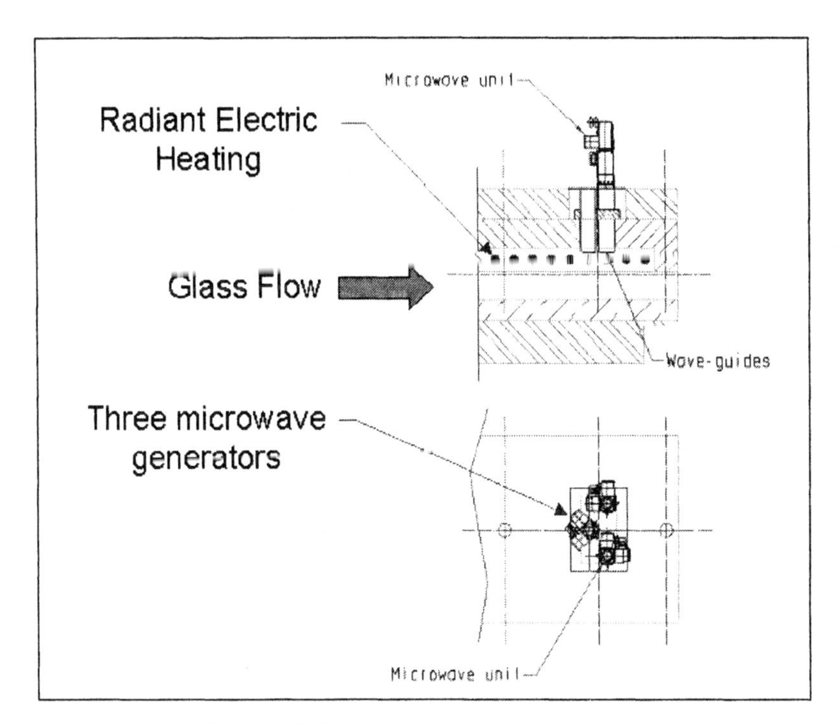

Figure 4. Barium glass installation.

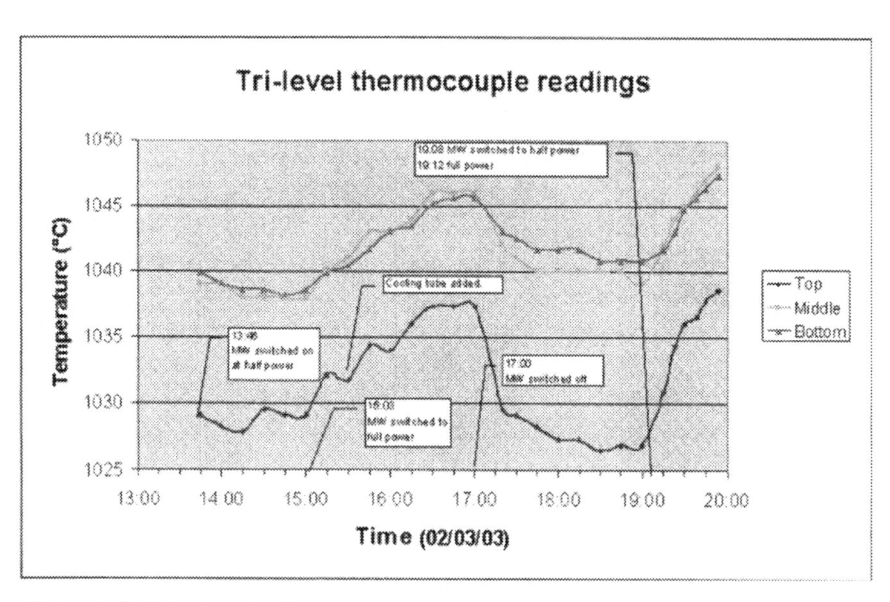

Figure 5. Barium glass results.

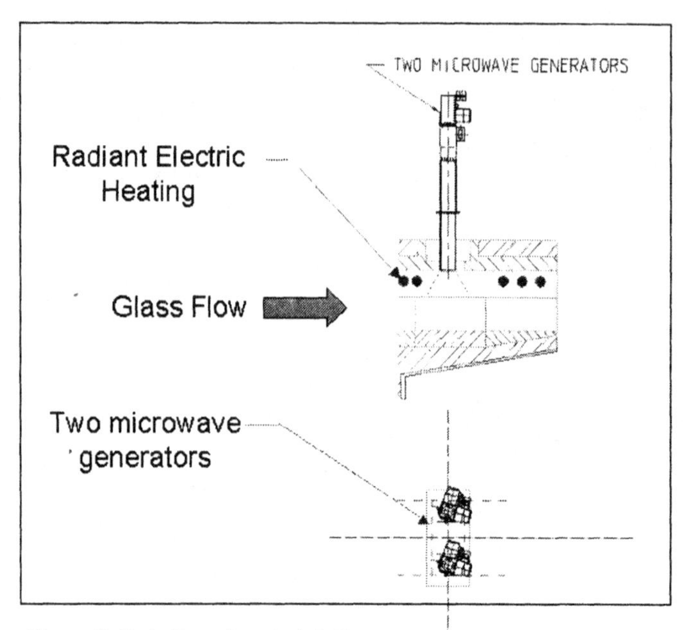

Figure 6. Soda-lime glass installation.

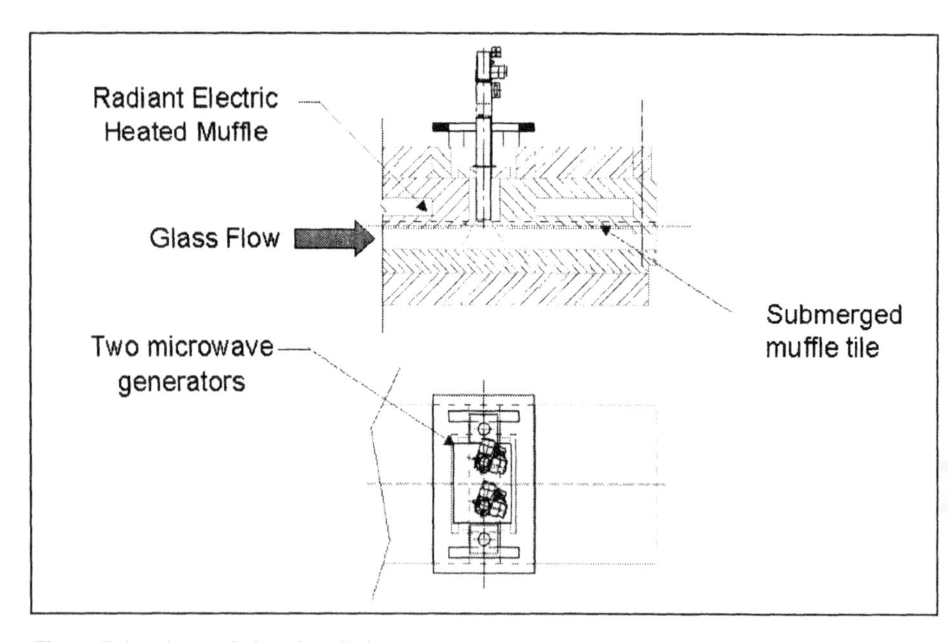

Figure 7. Lead crystal glass installation

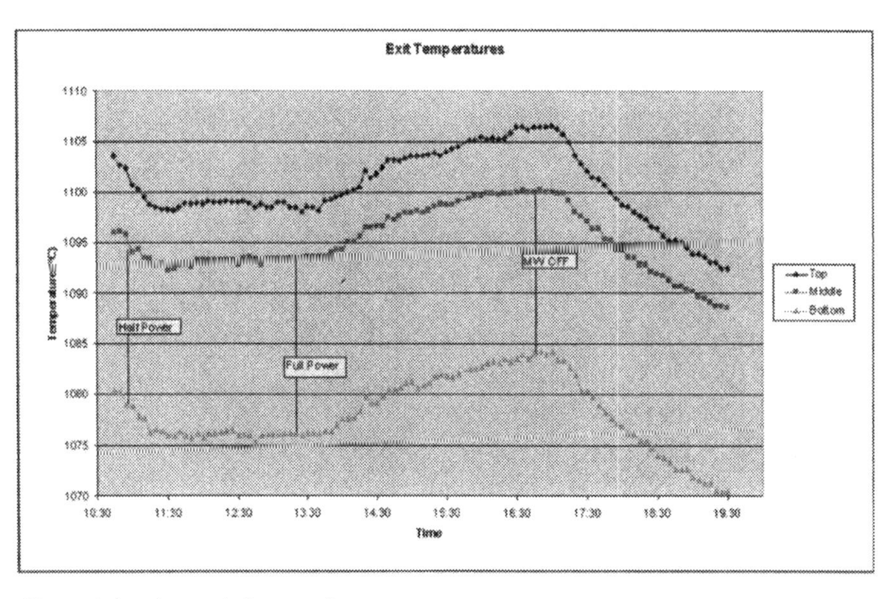

Figure 8. Lead crystal glass results.

Figure 9. Borosilicate installation.

The Development of The Emhart Glass 340 Forehearth

John McMinn, Emhart Glass, United Kingdom

Advances in forming machine technology and processes have been accompanied by similar, yet comparatively less heralded, advances in forehearth technology. For modern forming processes the rejection rate and the requirements for faster machine speed are inherently linked to thermal homogeneity levels in the gob and, by extension, to the efficiency of the forehearth conditioning system. The saying that "the bottle is made in the forehearth" is widely known, but if successful bottle making is an inherent feature of the forehearth, so also is the production of defective bottles.

Chipped finish	Sunken sides	Split finish	Heavy bottom
Broken finish	Bulged sides	Check finish	Rocker bottom
Over press	Wash boards	Crizzled finish	Swung baffle
Corkage check	Cold moulds	Finish checks	Wedge bottom
Out of round	Flanged bottom	Unfilled finish	Brush marks
Finish tear	Thin Bottom	Choked neck	Shear marks
Bent finish	Stuck ware	Bottom check	Broken ware
Bent neck	Heavy bottom	Thin ware	Loading marks
Hollow finish	Rocker bottom	Spikes	Glass distribution
Neck tear	Swung baffle	Offset finish	Drag marks
Shoulder check	Wedge bottom	Bulged finish	Lap marks
Thin shoulder	Brush marks	Pressure check	Out of shape
Sunken shoulder	Shear marks	Blank mould seam	Loading marks
Panel check			

The Emhart book of glass container defects lists some 63 ware faults of which 53 can be attributed to the forehearth. Although many of these defects are multi-causal, the potential impact on pack-rate from a poorly designed and inefficient forehearth is clear. Forehearth induced container defects are usually associated with glass thermal homogeneity and temperature stability and these in turn are determined by the efficiency of the forehearth and its associated subsystems.

Pack rate efficiency is also influenced by the speed with which the forehearth can respond to changes in the forehearth environment, for example caused by tonnage changes on an adjacent forehearth, and by the time required to produce stable glass after a job change. Again forehearth response time is determined by the efficiency of the forehearth and its associated subsystems.

Forehearth Design and Heat Transfer

Heat transfer processes in forehearths are very complex. Glass conditioning is characterized by banded internal radiation, natural convection, and thermal coupling of combustion gas, refractory, and glass. However a key element in the conditioning process is the forehearth roof block. The profile of the roof block and the number, geometry, position and control of the forehearth exhausts are vital elements in forehearth operation.

The relative glass and roof block surface temperature is also a key element in the radiation process between glass and refractory and is governed by the Stefan-Boltzman law

$$Q = \varepsilon \sigma A (T1^4 - T2^4)$$

Where ε (emissivity), σ (Stefan-Boltzman constant) and A (area) are constants. T1 and T2 are the glass and refractory surface temperatures.

The amount of radiation absorbed into the glass or radiated from the glass to the roof block surface is therefore dependent on the relative temperature of the glass and roof block surfaces. The temperature of the roof block is a controlling factor in determining the temperature of the glass stream, and the ability of the forehearth to selectively control the roof block temperature provides a powerful means to achieve thermal conditioning of the glass.

The roof block temperature is influenced by a variety of factors.

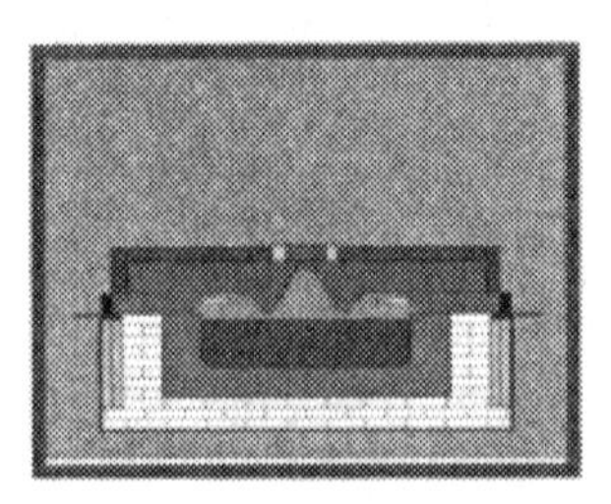

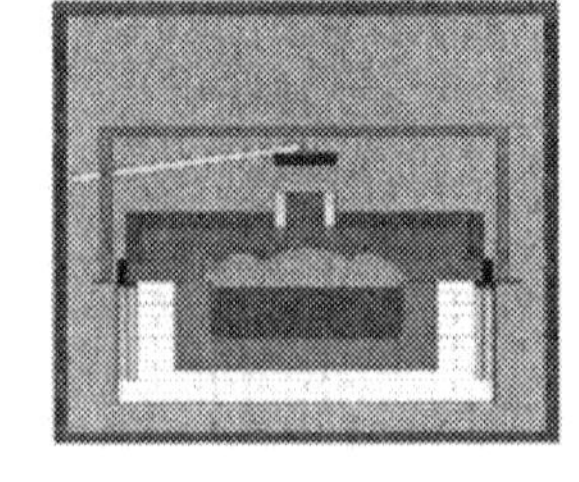

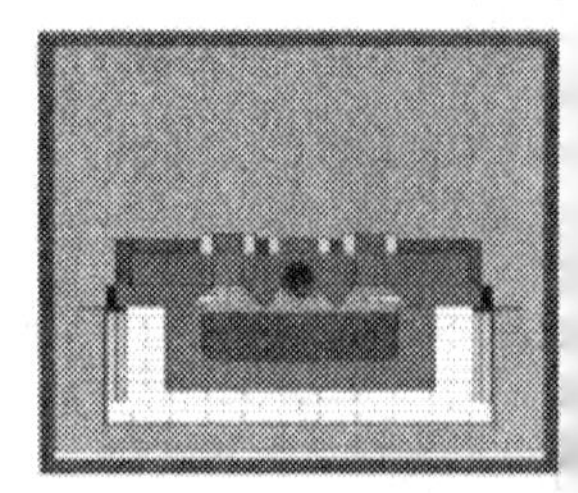

Figure1.

The shape or profile of the roof block is an important element in forehearth control; so much so that individual forehearth designs are characterised and can be identified by the shape of the roof block. The profile should assist both the cooling and heating functions of the forehearth. The flue exhaust geometry is also of key importance. The number, geometry and positioning of the flue exhausts can influence the flow of both cooling air and combustion gases within the forehearth and consequently influences the roof block temperature. Accurate control of these flues is essential. Finally and obviously, the method of cooling and the control of the cooling process greatly influence the roof block temperature.

Modern forehearth designs display a variety of cooling methods including radiation cooling, direct air cooling and muffle cooling. All three methods have advantages and disadvantages, proponents and opponents but, irrespective of cooling method, the process of cooling is vital and twofold. Firstly the cooling system must accommodate the glass load requirements to reduce the bulk glass temperature to that required by the process. The second, and crucially important function, is to provide thermal conditioning of the glass stream. It is the symbiotic relationship between roof block profile, flue geometry and flue control that provides an effective glass conditioning system.

The majority of commercial forehearth designs employ some form of forced air convection cooling. These are classed as direct cooling and indirect cooling. Direct cooling systems (usually but not exclusively) introduce cooling air into the forehearth chamber at the beginning of the zone forcing it longitudinally in the direction of glass flow to be exhausted at the end of the zone.

Indirect cooled forehearths operate using heat transfer plates embedded in the forehearth superstructure above the central glass stream. The plates are covered by a longitudinal refractory muffle extending over the length of the cooling zone. In operation, the central glass stream radiates heat to the cooling plates, which in turn are cooled by the airflow passing through the muffle. By automatically controlling the flow of the cooling air through the muffle, the amount of heat removed from the glass can be controlled.

Each system has unique advantages and disadvantages. Direct cooling provides faster response and greater heat removal than indirect cooling. Muffle cooled systems have the disadvantage that the effective cooling area is the sum of the area of the cooling plates, not the full length of the cooling zone. The main advantage of muffle cooled forehearths is that there is no interaction between the heating and cooling components due to their physical separation.

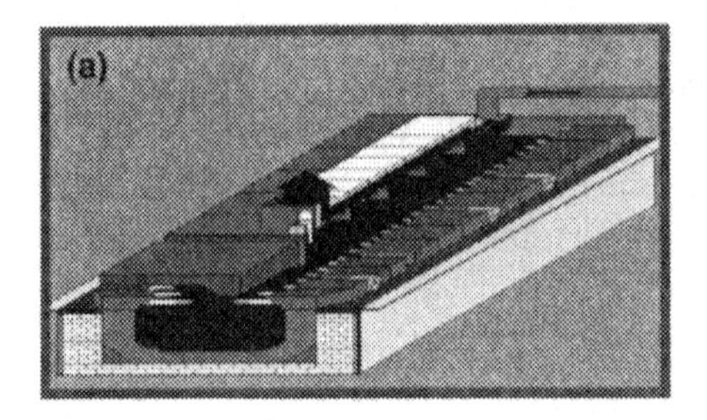

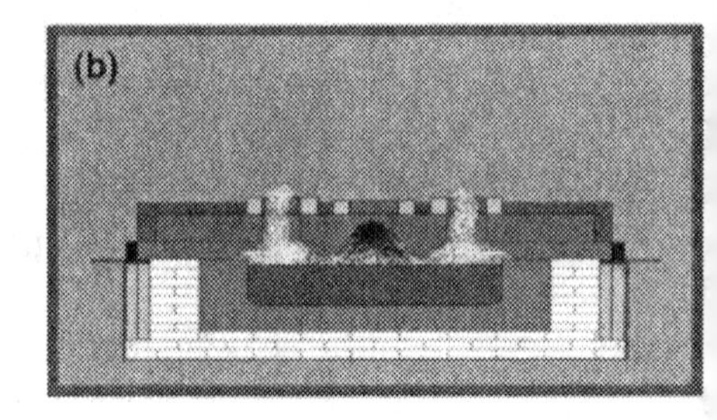

Figure 2. (a) Indirect cooling. (b) Direct cooling.

For both systems however the amount of heat a particular cooling zone can remove, and the speed of response, is finite. The effectiveness of the indirect system is limited by the number of cooling plates which can be accommodated in the forehearth superstructure. The direct cooled system is limited by the amount of air that can be introduced into the forehearth chamber without interfering with the combustion gases and without adversely affecting the surface layers of the glass stream.

The effectiveness of both systems can however be extended to produce a forehearth system capable of greater glass loads and able to provide faster forehearth response with higher thermal conditioning. This can be achieved by combining both cooling methods in a forehearth system equipped with a flue control system capable of controlling both cooling air flow and combustion gas flow within the forehearth chamber. This is the basis for the Emhart Glass 340 forehearth system.

The interactions within the forehearth chamber and the influence of the superstructure and exhaust flues are very complex. Extensive use of mathematical modeling was made to determine the number of flues required per zone, their size and geometry, their position in the refractory superstructure, maximum cooling air flow rates and flue control.

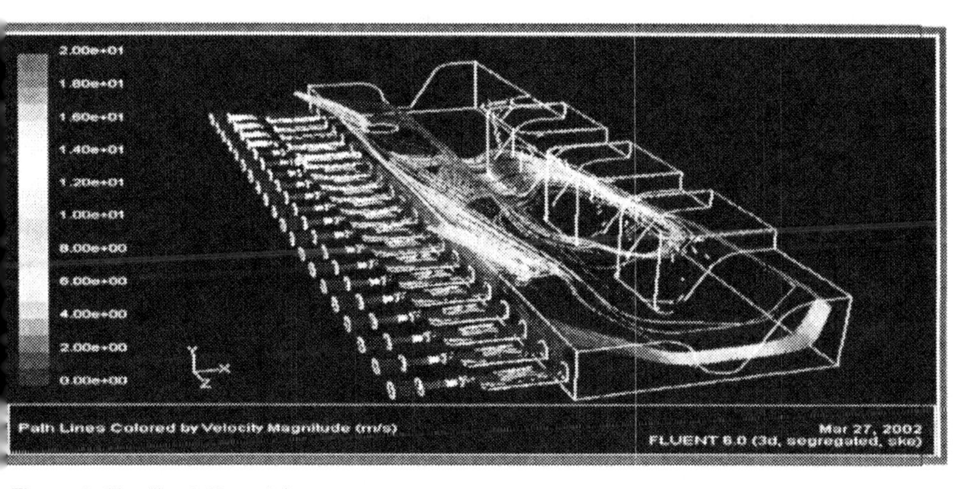

Figure 3. Five-flue 340 model.

The 340 employ a unique simultaneous dual cooling system in which a combination of muffle cooling and direct forced convection cooling is used.

Each cooling zone is configured with 4 automatically controlled side exhaust flues, an automatically controlled direct cooling air exhaust flue positioned over the central glass stream at the end of the zone, and a static muffle cooling exhaust flue positioned adjacent to the direct cooling flue.

Control of the five-flue automatic exhausts effectively determines the flow of cooling air and combustion gases both longitudinally and laterally within the forehearth chamber. This in turn determines the relative surface temperature of the forehearth roof blocks and the glass stream, and consequently provides a powerful mechanism to selectively input or remove heat from the glass ensuring tight control of the thermal conditioning process.

In addition to the thermal conditioning benefits, this configuration provides the forehearth with an unprecedented degree of cooling power. However the high cooling capacity is not merely to cater for large tonnage production. The ability to remove large quantities of heat from the glass allows much higher grades of insulation to be used. Higher substructure and superstructure insulation levels result in reduced structural heat losses that in turn result in significantly improved conditioning when operating at lower tonnages. The increase in cooling capacity can be exploited to increase the tonnage range over which an individual forehearth can operate. The same argument is also applicable for extending gob temperature and entry temperature ranges.

Mathematical modeling proved the importance of flue control in the glass conditioning process. Consequently the 340 design incorporates a unique damper/flue combination which provides a high degree of controllability of the forehearth exhaust gases and cooling air. In this arrangement, each damper and its associated flue are in contact throughout the operational range. Although each zone damper control actuator is required to move five dampers, the torque requirement has been minimised by the design of the flue block and no counterbalance system is required. The damper and flue blocks are made from different Zircon Mullite materials to ensure preferential wear rates between damper and flue.

It is predicted that it will not be necessary to replace the flue blocks for the duration of the campaign and, depending on operational considerations, the damper blocks may require one change throughout the campaign.

Figure 4. 340 damper control mechanism.

The 340 contact damper/flue arrangement also incorporates a self-cleaning facility. As the damper block travels across the flue block, any condensates that have built up on the hot face of the damper are removed by abrasion. These condensates fall into a special chamber in the flue block and can be easily and quickly removed through an access plug in the outer flue block wall.

Operational results of Emhart Glass 340

A main conclusion of the mathematical model was that the 340 forehearth would significantly increase both the tonnage range and the gob temperature range for a given forehearth dimension. Emhart Glass employ several methods for determining the required forehearth dimensions based on glass colour, tonnage and required temperature drop. One of these is illustrated below. The two curves represent the upper and lower limits of a 36 inch wide 540 forehearth configured with 18ft cooling. The green and red squares represent the results of an operational 340 forehearth configured with identical dimensions. The extended range of the 340 is evident from the graph.

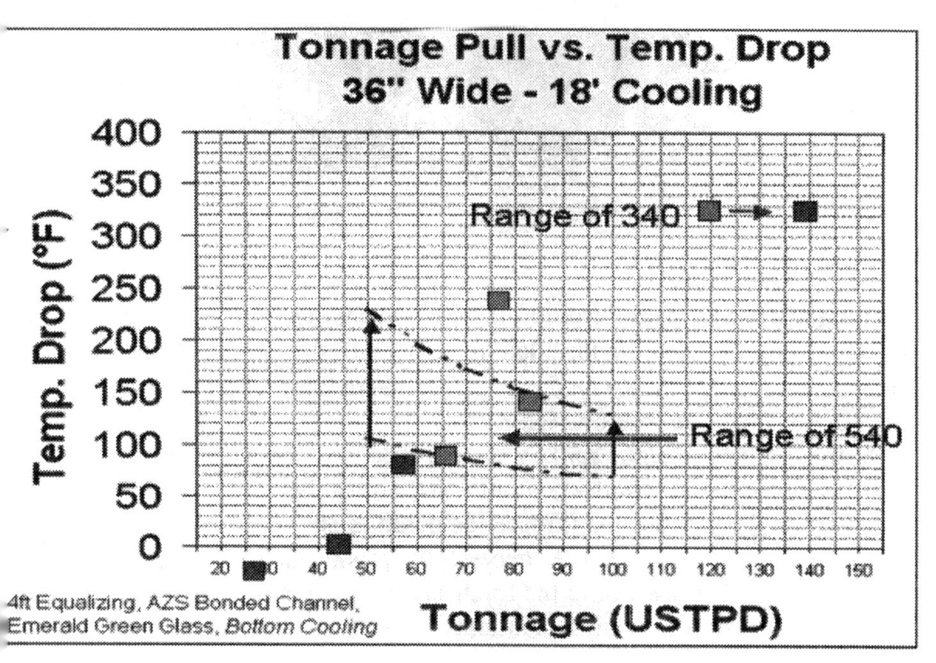

Figure 5. The 340 has been sold into the USA, South America, the Middle East and will shortly be installed in Europe. The first 340 forehearth was commissioned in January 2002 and so far has produced bottles with a gob weight range of 85g to 1,382g of green glass – a gob ratio greater than 1/16.

Figure 6. 85g sauce 918gWine Flask 1382gWine Flagon.

The first container produced on the 340 was a green, 918g green flat-sided wine flask. Previous to the installation of the 340, this job had been associated with a relatively low pack rate due to glass temperature related problems. Using the 340, the pack rate rose by 7%.

On the same forehearth, an 85g sauce bottle was produced. To manufacture this bottle prior to the introduction of the 340, it had been necessary to lower the furnace level, to increase the furnace temperature and to insulate the forehearth with insulation wool. The 340 successfully produced this bottle without any adjustment to the furnace or to the forehearth insulation levels.

The second 340 forehearth was installed in Mexico. This line controlled by the new Emhart Glass HWC900 SCADA control system produces green and flint ware.

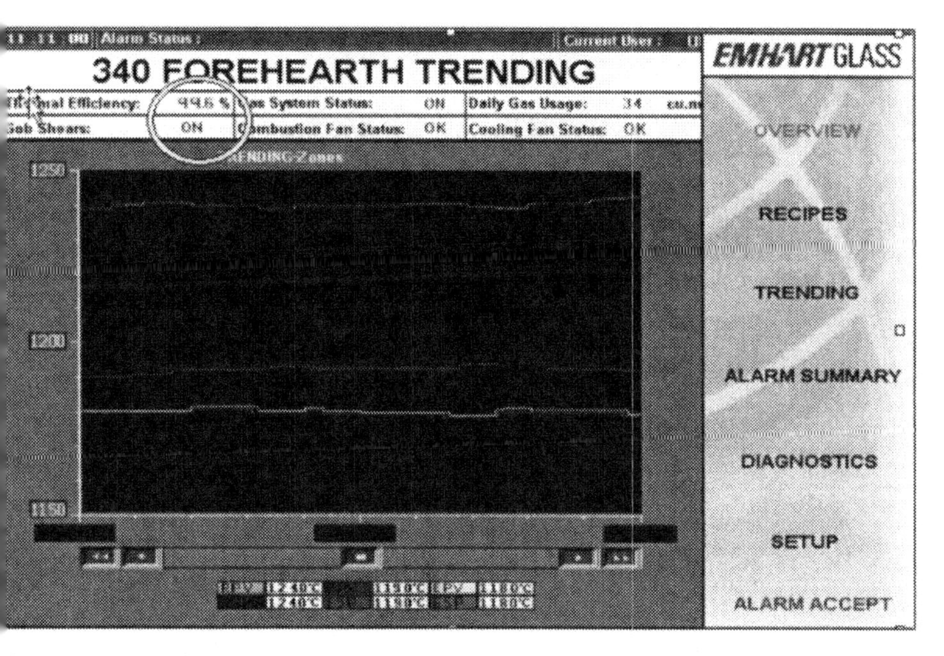

Figure 7. This 340 has consistently produced thermal efficiency values in excess of 99% on both green and flint containers.

Conclusion

The Emhart Glass 340 forehearth is a unique glass conditioning system combining speed of response with extended gob temperature and tonnage ranges. The system has attracted further sales into the USA and the Middle East and is due to be launched into Europe. The future of this system is as clear as the glass it provides.

New Developments in Stirrer Technology

Duncan R Coupland and Paul Williams,
Johnson Matthey Noble Metals, United Kingdom

Abstract

Stirrer design has undergone only minor change in the last 2 to 3 decades. The stirrers, stirrer bars, blenders, homogenisers, screw plungers and plunging stirrers used in numerous glass processing applications tend to be effective in their operation, well proven and of predictable life-span. Until recently there has been little real incentive to develop the technology of stirring.

However, the current climate of the glass industry in general has resulted in powerful drives towards cost reduction and improved operational efficiency, recently with particular attention to platinum inventories.

The challenging operational environment of most stirrer applications has necessitated robust engineering of the stirring devices, often involving large quantities of platinum or platinum alloys. As a result, the amount of platinum required for stirring applications can often be a significant portion of the total platinum alloy inventory of a glass company.

In response to the desire to reduce platinum inventories, developments in stirrer technology have been able to reduce the platinum requirement (in some cases by >90%) without jeopardising the stirring effectiveness or stirrer longevity.

This paper briefly reviews the benefits of the different types of stirrers available to glass makers and introduces a new concept in stirrer design that offers great potential benefits. Recent trials data for this technology "The Diffusion Choke" clad stirrer, is presented and the implications discussed.

Introduction

Glass homogeneity is one of the pre –requisites of good quality glass. Over many years glass melting furnace designs have been developed to achieve high degrees of inherent mixing and so to deliver uniform glass into the forehearth. However, the necessity to condition glass flowing towards the working end can negate some of this design and give rise to thermal and compositional inhomogeneities that could compromise finished product quality. The solution is forehearth stirring, which is widely employed. The introduction of stirrers may be a double-edged "sword", with upside in respect of stirring but downside in respect of their physical presence.

The choice of material for the stirrer is key to gaining optimum benefit, and as with all things, has to be a balance of cost, effectiveness, and durability. All of these are aspects that reflect the nature and application of the glass itself; notably viscosity, temperature, corrosivity, quality and value. Stirrers with glass contact surfaces made in platinum or its alloys present the ultimate solution to this issue, but for many applications the cost has been historically too high. This changed with the introduction by Johnson Matthey of ACT™ platinum coated ceramics by which immense improvements in glass resistance have been achieved, at relatively moderate cost by virtue of the much enhanced durability and longevity as compared with unprotected ceramic Ref 1. However, for those glasses of extreme value, specifically optical glasses where quality and clarity are paramount, stirrers fabricated from platinum alloys have been used, although having limited durability in high viscosity glass, especially for large components. Molybdenum cores were found to be capable of providing the strength required for parts such as gobbing and high energy stirrers, although extreme measures are required to avoid it from literally going up in smoke in air at any temperature above ca 400°C (750°F).

Platinum alloy coated, high strength oxide dispersion strengthened superalloys, have been trialed over the years, but without significant success, until now, with the introduction of the Johnson Matthey diffusion choke technology. This paper is focused on this new development.

ACT™ Platinum Coated Stirrers

The ACT™ platinum coating technology has now been in molten glass for ten years. Some of the earliest applications were coated ceramic stirrers for application in the severest conditions such as opal and borosilicate glass and colouring forehearths. The objective was to stop the ceramics from eroding and allow continuous efficient stirring to occur and the effectiveness can be seen in Figure 1. Having proven the concept the technology is now being used to re-define the nature of stirrers and to gain further advantage over the conventional, in this case, conventional fabrication, Figure 2. In a recent application, co-planar ceramic stirrers with ACT™ platinum alloy coatings, Figure 3, have replaced helical stirrers fully fabricated from platinum alloy. The performance in stirring TV panel glass has dramatically assisted the economics in this difficult area, by reducing the required platinum content from a total of 84kg to only 8kgs. This is partially accomplished by the superior design of the stirrer resulting in a smaller number being required, i.e. from 10 to 4, and by the reduced thickness of the coating as compared to the cladding.

The design of the ACT™ coated stirrer is dictated by the ceramic selected and the requirements of the application, and many different configurations have been defined and utilized.

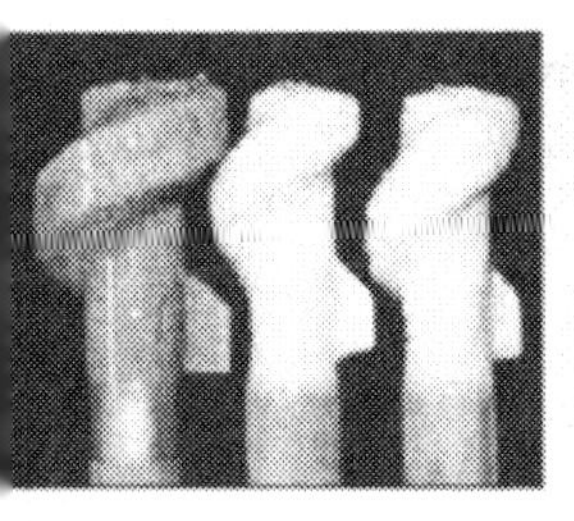

Figure 1. The benefit of ACT™ coating in a colourine forehearth.

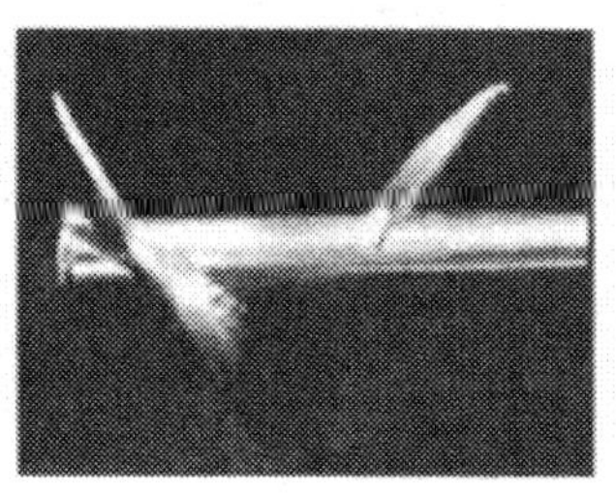

Figure 2. Conventional fabricated stirrer.

Figure 3. ACT™ coated co-planar stirrer.

Platinum Clad Base Metal

Molybdenum has for many years been selected as the material of choice for structural applications within the glass furnace, and it is well known to perform well in molten glass. However it has a major limitation in that although it is used extensively as electrodes in electrically heated furnaces, if free oxygen impinges on the hot surface it burns rapidly. To be effective the molybdenum must be protected if it is to function at any temperature greater than about 400°C (750°F). Therefore in its major application, as resistance heating electrodes it must be water cooled to ensure the zone not protected by immersion in molten glass is kept below this critical temperature. Platinum can perform without this limitation and can be successfully used in similar applications without recourse to water-cooling. Convention assumes that the platinum would make the electrode too expensive for use, but this is not always the case and the introduction of the ACT™ platinum coating technology has allowed designs to be generated that have all the advantages of platinum without the disadvantages.

In glass stirrer technology the strength of molybdenum is desirable for critical applications where the shear strength requirements are very high and where unexpected failure would be expensive, but environmental protection is critical to achieve this. Platinum cladding has conventionally been utilised in a simplistic symbiosis, whereby the platinum alloy cladding protects the molybdenum that provides the strength. As in many symbiotic relationships there is also a parasitic component and the two materials can, under some circumstances, interact with the formation of potentially detrimental intermetallic phases (Ref 2). The effect of this can be seen in Figure 4 which shows the weight change observed

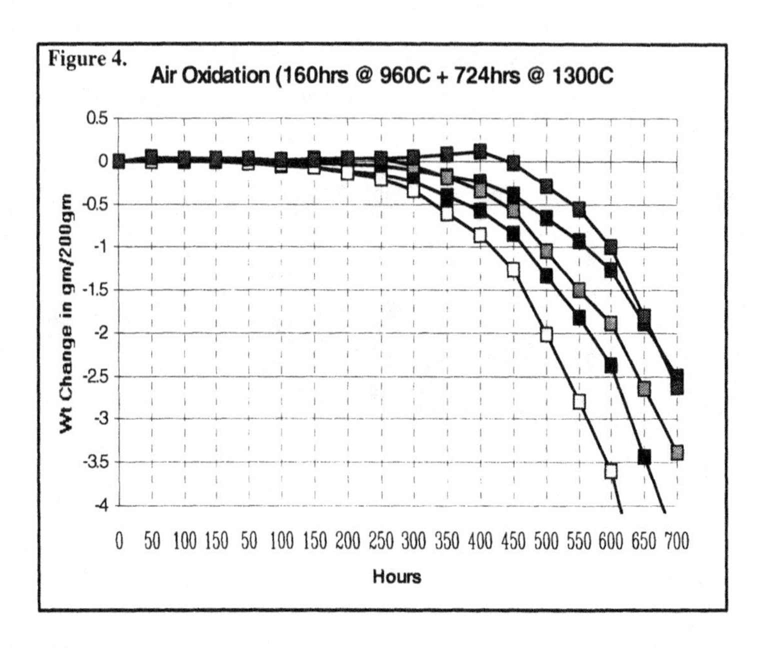

Figure 4.

for a series of molybdenum samples protected by high thickness platinum coatings that have been demonstrated to protect ceramic substrates so effectively. Essentially the simple platinum layer can protect the substrate until interdiffusion and interaction promote failure of the platinum layer. Once this has happened rapid oxidation of the molybdenum occurs with the dramatic loss of weight as shown. The use of a ceramic barrier layer to keep the two metals apart is a natural progression. However, this ceramic barrier can only control interdiffusion and oxygen removal from the space is essential (Ref 3). This situation must be maintained for the duration of the component' service life.

Such design of stirrers has been used very successfully for several decades, and with care can give lives of more than five years. However, when the cladding fails either by mechanical damage, physical change or chemical attack, the introduction of oxygen onto the molybdenum can cause a dramatic and rapid failure. Often this failure can be anticipated and avoided, but if it happens unexpectedly the damage to the forehearth and the down time resulting can be extreme.

It has always been an objective to find an alternative to molybdenum as the core material, and trials of various materials have been undertaken. The class of alloy most likely to be suited to this arduous task is the superalloys. These materials were developed specifically for the gas turbine industry and were designed to have excellent strength, combined with very good oxidation resistance up to temperatures of approximately 1100°C (2000°F), although oxide dispersion strengthened alloys are available where useable strength is still available at 1300°C (2370°F). Some of these alloys have considerable resistance to the atmosphere above molten glass and indeed even when submerged. However, they tend to erode relatively rapidly at the glass line, causing both structural weakening and potential glass colouration problems. Although it would seem feasible to use a platinum cladding to negate this weakness, work done a few years ago Ref 4 showed that the tendency of nickel and platinum to interdiffuse was too great for long term success. An example of an oxide dispersion strengthened alloy of this genre, which was been ACT™ platinum coated and then tested in molten TV glass for 300 hours at 1150°C (2100°F) is shown in Figure 5. Approximately the first quarter of this component had platinum deposited directly onto the base metal substrate and through diffusion has resulted in the development of surface oxide on top of the platinum. However the lower 3/4 of the sample was provided with a ceramic interlayer which very effectively blocked the diffusion, although an additional slight colouration of the glass still attached to the sample surface hints at a finite level of migration of iron, nickel or chromium from the core alloy. Therefore, although the ACT™ coating technology would offer an improvement, a further technological advance would be required to allow the effective use of the ODS materials. A novel method has been defined, and patented (Ref 5), and the next part of this paper will describe the design and performance of a stirrer made using this technology.

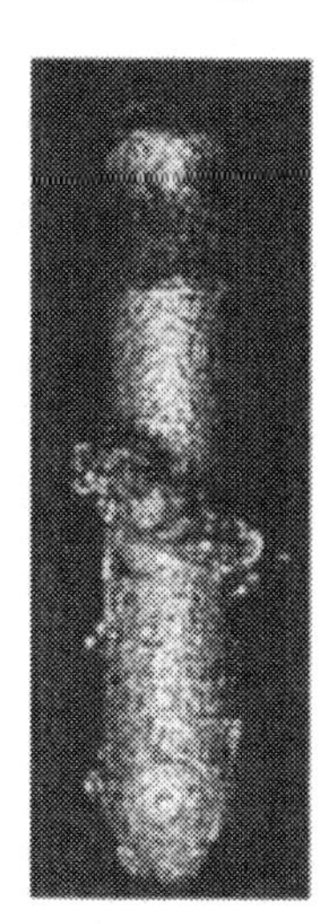

Figure 5. ACT™ platinum coated oxide dispersion strengthened alloy.

The Diffusion Choke

It is clear from the previous sections that companies melting glass and forming it into high quality shapes often require stirrers capable of operating reliably for long periods in the temperature range 1000°C (1830°F) to 1300°C (2370°F). Technical solutions exist but all have limitations either in performance or cost, whether this be the inherent limitations in design embodied in the excellent ACT™ coated ceramic stirrers, or the cost and inherent potential for catastrophic failure of the extremely strong platinum clad molybdenum ones. The "diffusion choke" was designed to provide a viable alternative to this latter, whereby the risk of catastrophic failure is eliminated allowing for some potential reduction in required platinum cladding thickness, and hence some modest cost reduction.

The technology is exemplified by a recent trial conducted in TV glass using a PM2000 stirrer shaft and a 20%Rh-Pt sheet metal fabrication or cladding. Between the two was placed a mesh of finely knitted 10%Rh-Pt alloy; the "diffusion choke", Figure 6. It is designed to separate the cladding from the substrate and so reduce the diffusion arising from contact at high temperature that is known to happen and which is exemplified by the laboratory test sample shown in Figure 5. Its second function is to maintain an airway to the outside, and so to ensure an adequate supply of oxygen to the surface of the core alloy. This ensures that the inherent oxidation properties of the alloy are developed and maintained throughout prolonged operation at elevated temperatures. The "choke" does restrict oxygen flow to the alloy surface, and ensures that excessive oxide thickness cannot develop which under certain conditions might otherwise give rise to an aggressive form of rapid oxidation. An interesting additional advantage of the presence of platinum in contact with the core alloy is the known effect on increasing oxide stability (Ref 6). The design for the stirrer is shown in Figure 7.

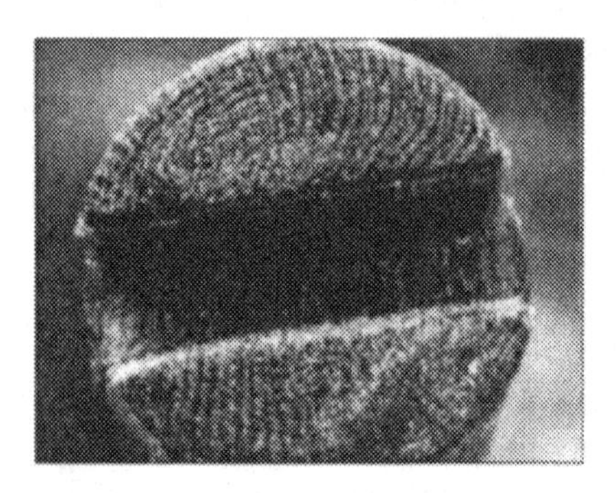

Figure 6. 10%RH-Pt Gauze wrapping on the core.

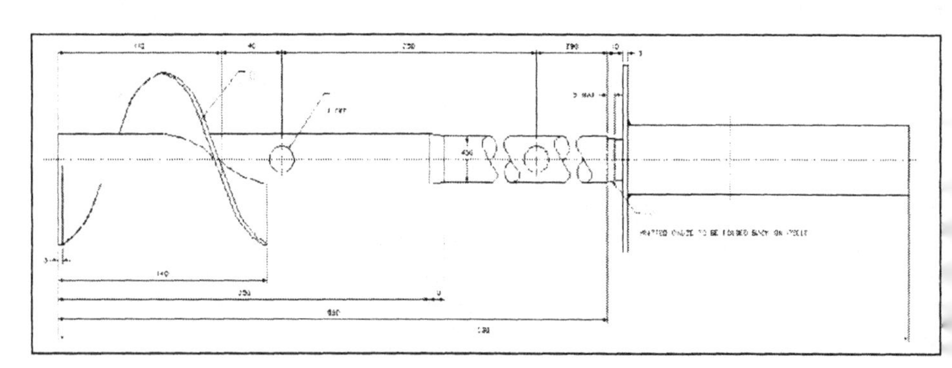

Figure 7. Schematic of helical bladed stirrer incorporating the "Diffusion Choke".

The service trial was extended from the original that had been planned to last only 6 months, to 20 months as it had progressed so well. Even at 20 months its removal was only undertaken with great reluctance, based on the old adage; "if

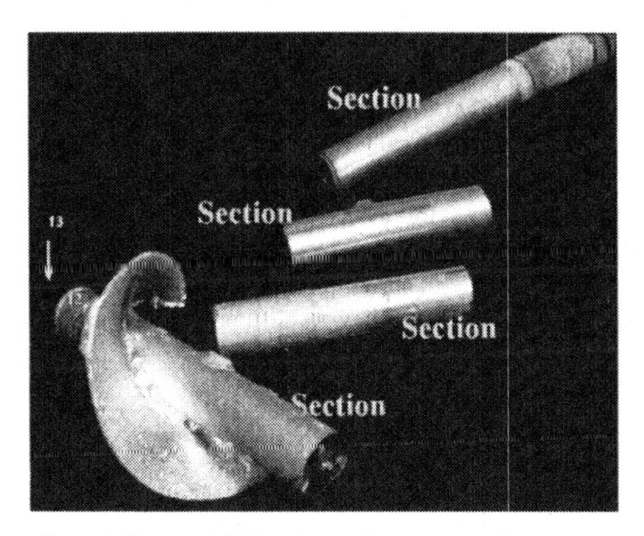

Figure 8. Returned Stirrer after preliminary sectioning

it ain't broke don't fix it". The removed stirrer was returned to Johnson Matthey Noble Metals where it was given a full investigation. Initial visual examinations were very positive, with no indications of degradation of surface at any point along its length being observed. The items can be seen after preparation for analysis, in Figure 8, with the analytical sample points identified numerically. The discolouration at the upper end of the shaft was crusty and mineralised, indicative of deposition from the vapours of the furnace, and at the glass line the platinum alloy was slightly brighter possibly indicating some minor surface interaction. A microfocus XRF unit was used to examine the composition of the external surface, along its length. The results are presented in Table 1.

Table 1. Microfocus X-Ray Fluorescence Analysis of the Stirrer Surface

Sample	Pt	Rh	Fe	Cr	K	Ca	Sb	As	Bi	Ba
1	27.4%	19.7%		0.7%	44.1%		7.5%	0.6%		
2	14.5%	27.7%	0.2%		5.2%	21.3%	31.0%			0.1%
3	77.2%	17.6%	1.1%		3.4%				0.7%	
4	77.9%	18.8%			2.5%				0.7%	
5	79.7%	19.3%			1.0%					
6	80.2%	19.2%			0.6%					
7	80.4%	19.6%								
8	80.1%	19.5%								0.3%
9	80.1%	19.9%								
10	80.3%	19.7%								
11	80.4%	19.6%								
12	80.9%	19.1%								
13	80.7%	19.3%								

The analyses of the upper region of the stirrer, positions 1-6, show that the yellowish encrustation has derived from the molten glass probably via condensation from the gas phase. This is of course normal and expected. The rhodium contents of the alloy are exactly as the original alloy specification, within the error expected for the analytical equipment. On the lower portion of the stirrer there were no "foreign" elements detected on the alloy surface except for a trace of barium, a glass component, in the region of the glass line. This lack of any surface impurities after immersion in molten glass indicates that the component had been relatively thoroughly cleaned before being returned for investigation, and therefore the lack of evidence of through diffusion from the substrate was, at this stage still unproven. Further disassembly of the stirrer was therefore required.

Table 2. XRF Results for the Inner Surface of Rh-Pt Tube.

Sample	Pt	Rh	Fe	Cr
1i	79.8%	20.2%		
2i	79.1%	20.9%		
3i	83.7%	16.3%		
4i	83.4%	16.6%		
5i	83.4%	16.6%		
6i	82.5%	17.5%		
8i	82.7%	17.3%		
9i	82.6%	17.4%		
10i	82.4%	17.6%		
11i	82.9%	16.9%		0.2%

The cladding on Section A slid easily from the base metal shaft. The diffusion choke was fixed onto the 20%Rh-Pt tube, but further examination showed this to be a very slight adhesion and minimal tension was required to allow the gauze to be removed. In fact the same situation was found for the whole length of shaft, except of course where the fixing screws had been securely positioned to transfer torque on the shaft to the cladding and hence the stirrer blades themselves. Along most of the length of the shaft the gauze had remained shiny and metallic, but in one region there was some greyness and at the very top some gauze was yellow/brown. Analyses of gauze samples and of the inside of the tube at points corresponding to the external analyses. These results are presented in Tables 2 & 3 respectively.

The analysis of the inner surfaces of the 20%Rh-Pt protective tubing, with the possible exception of one sample below the glass line, showed no elements to be present that would not have been present in the original alloy. Interestingly the

Table 3a. Analysis on the outer surface of the Diffusion Choke.						
Sample	**Pt**	**Rh**	**Fe**	**Cr**	**Y**	**Ti**
1g	88.3%	11.7%				
6g	87.8%	12.2%				
8g	86.1%	13.9%				
11+g	85.8%	14.1%		0.1%		

Table 3b. Analysis on the inside surface of the Diffusion Choke.						
Sample	**Pt**	**Rh**	**Fe**	**Cr**	**Y**	**Ti**
1gi	89.6%	10.4%				
6gi	87.7%	11.2%	0.3%	0.1%		0.7%
8gi	85.7%	13.7%	0.2%	0.4%		
11+gi	85.9%	13.0%	0.2%	0.5%		0.3%

rhodium level in the inside surface of the tube shows an approximate 3% reduction from the original bulk alloy composition, and indeed the extant, average outer surface composition. The question as to where this had gone was answered by the analysis of the Diffusion Choke itself. This showed a corresponding increase in rhodium content indicating that either a diffusion process or a vapour phase transfer process had been operating. In addition to an increased average rhodium content the diffusion choke exhibited measurable levels of iron, chromium and titanium on the side in contact with the base metal substrate, but almost none on the side in contact with the Rh-Pt tube. The surface of the choke in contact with the core was slightly discoloured, and appeared to have a physical contamination rather than a chemically bonded one.

The core used was an iron-based alloy with major additions of chromium and aluminium plus various other minor ones. The key to its high strength at elevated temperature is the presence of yttria which as a dispersed, stable oxide, provide grain boundary strengthening. Table 4 shows the results for microfocus XRF of the surface of the alloy after service, although the absence of values for aluminium is an artefact of the analytical technique, rather than a mechanistic issue. Alternate analytical techniques can be used to confirm that aluminium has been retained.

Table 4. Microfocus XRF analysis of the Substrate Core Surface.						
Sample	**Pt**	**Rh**	**Fe**	**Cr**	**Y**	**Ti**
14	1.0%	0.1%	79.1%	17.8%	0.8%	1.1%
15	5.8%	0.5%	73.6%	18.2%	0.8%	1.1%
16a (light)	6.1%	0.5%	72.9%	18.5%	0.7%	1.2%
16b (dark)	2.5%	0.3%	75.1%	18.5%	0.7%	2.9%
17	22.6%	1.7%	56.3%	17.7%	0.7%	1.0%
18	7.4%	0.7%	72.7%	17.1%	1.1%	0.9%

The presence of the occasional high values for platinum and rhodium on the surface of the base metal core was due to very small adhered flecks of platinum. At this time the nature of these tiny platinum rich particulate has not been determined, so that it is impossible to say whether they have been transported by a vapour phase mechanism or are simple physical artefacts. Visual observation, however, clearly indicated that a thin, protective surface oxide had been formed, which would be anticipated to change only slowly allowing protection to the substrate for a very long time.

Conclusion

The in-service trial and subsequent destructive analysis of the 20%Rh-Pt clad, ODS iron based alloy, stirrer reported in this paper, has demonstrated that a new technology has arisen to overcome many of the problems associated with traditional clad molybdenum stirrers. The technology offers a breakthrough in stirrer design and as a result an additional tool to the glassmaker in his use of stirrers for improving his glass quality. In this trial the stirrer design was simple, with much reliance on traditional fabrication skills in its construction. The Patented diffusion choke technology does have the potential for successful use in a wider range of stirrer types, and perhaps additional applications, where high strength, durability and longevity, without risk of catastrophic failure, are paramount.

Overall the diffusion choke has maintained an effective barrier to degradation of the stirrer for the 20 months of service. Indeed the evidence of the analyses suggests that the component would have maintained integrity for a much longer time. However, the original objective of the analysis was to allow a prediction of probable lifetime, but the lack of significant changes means that there are no indicators to measure adequately. It must therefore be assumed that the component would have been useable for a very much longer period, perhaps two, three or even four times as long. Such lifetimes, of the order of 5-10 years must be considered as making this design of stirrer both a technical success, and potentially a commercial opportunity.

BIBLIOGRAPHY

1. Coupland et al, Glass Technology Vol 37 No:4 Aug 1996
2. Selman; PMR Vol 11No:4 October 1967
3. Darling & Selman PMR Vol 12 No: 3 July 1968
4. Johnson Matthey Noble Metals Internal communication
5. "Platinum Metal Based Article for High Temperature Applications" WO 03/059826 A1
6. "Platinum-Enriched Superalloys", Corti et al; PMR Vol No: Vol 24 No: 1 January 1980

Spinel Refractories and Glass Melting

Chris Windle, DSF Refractories and Minerals, United Kingdom

Introduction

With the inception of oxy-fuel technology, longer campaigns, demand for a lower defect threshold, and higher melting temperatures, the environment of the glass melter has challenged refractory superstructure performance.

This melter evolution has paved the way for a new generation of refractories based on magnesium aluminate spinel.

Spinel is inherently resistant to both alkali (Na, K or mixed) and boron oxide, which provides potential melter application for a wide range of glass compositions including E, C and lighting.

Spinel superstructure is now established for soda lime silica glass production, with the predicted minimal interaction with alkali laden atmospheres fully validated.

In addition to chemical resistance, spinel shows excellent thermo-mechanical stability coupled with a capability to withstand recorded temperatures of 2000°C.

Driving forces for adoption include product cost, ease of installation, unique performance, low maintenance, low defect potential, and reliability.

Available pressed or cast a wide variety of shapes to varying degrees of complexity are available.

Recent development has spawned a spinel with phenomenal thermal shock resistance (TSR spinel), and this product is being utilised in hot repairs.

Structure and Composition

Magnesium – aluminate normal spinel is a member of the oxide group of general formula AB_2O_4. The oxygen ions adopt face-centered cubic close packing, with di-valent ions A^{2+} on tetrahedral sites, and B^{3+} ions on octahedral.

Spinel products can be manufactured in a variety of ways dependent on the target application.

Superstructure spinel (S Spinel) is hydraulically or impact pressed and fired to provide direct bonding through in-situ reaction of selected alumina and magnesia fines. The fines encapsulate the coarser particles consequently the chemical and thermo-mechanical properties of the product are directly influenced by the bond components. For high alkali/oxy fuel environments the total composition should be as near stoichiometric as possible, although the fines are maintained slightly magnesia rich to avoid β- alumina formation and the associated expansion.

For complex items a cast spinel (Spinel C) is utilised. This product is manufactured through a no cement route which provides ultimate temperature performance;although a corollary of this is small residual corundum content.

This does not seem deleterious to the product especially when adopted for burner systems which provides a gaseous shield.

Recent development (Spinel C1) has resulted in a corundum free system.

Although superstructure spinel exhibits good thermal shock resistance;field trials established that this product could not be hot inserted for repair work. This subsequently led to the development of Spinel TSR;which has been successfully installed into a hot gas out-take of an oxy fuel melter.

These products are not generic but relate to products currently manufactured under the trade names, Frimax 7 (S Spinel), Frimax C (Spinel C1), and Frimax HI (Spinel TSR)

Key chemical constituents and the derived mineralogical analyses (where indicated) for the three spinels reviewed in this paper are given in table 1.

TABLE 1. KEY CHEMICAL CONSTITUENTS AND THE DERIVED MINERALOGICAL ANALYSES

	S Spinel	Spinel C	Spinel C1	Spinel TSR
Al_2O_3	71.9	78.9	77.7	65.6
MgO	28.0	20.4	21.7	26.7
CaO	0.19	0.28	0.22	0.15
SiO_2	0.06	0.11	0.09	0.08
Fe_2O_3	0.03	0.12	0.08	0.03
Na_2O	<0.03	0.17	<0.03	<0.03
ZrO_2				6.98
C/S Ratio of bond	2.85			
C/A Ratio of bond	0.18			
XRD phase analysis				
MA	Major	97.3	98	95
M	Minor			0.8
A	Not detected	2.7	Not detected	
PSZ				6.2

MA; spinel, M; periclase (magnesia), A; corundum (alumina), PSZ;partially stabilised zirconia

It is important to note that the superstructure spinel only contains a total impurity content of ~0.28% (CaO, SiO_2, Fe_2O_3, Na_2O), and this is essential to maintain thermo-mechanical properties.

Although the total impurity content is extremely low, the lime: silica (C/S) and lime: alumina (C/A) ratios of the fine matrix must be controlled to ensure optimum performance specifically in high temperature and load applications.

A high C/S ratio promotes intergranular bonding;however this also theoretically promotes low melting phases within the $MgO-Al_2O_3-CaO-SiO_2$ quaternary system.

To combat this silica is maintained below 0.1%, and of greater significance the C/A ratio is kept low promoting the high melting temperature calcium aluminates.

A back scattered image of the superstructure spinel microstructure reveals a very dense texture with isolated porosity and minimal glass phase between the grains (2 microns thick). This structure not only provides exceptional alkali resistance, but also excellent resistance to deformation under load (Fig.1).

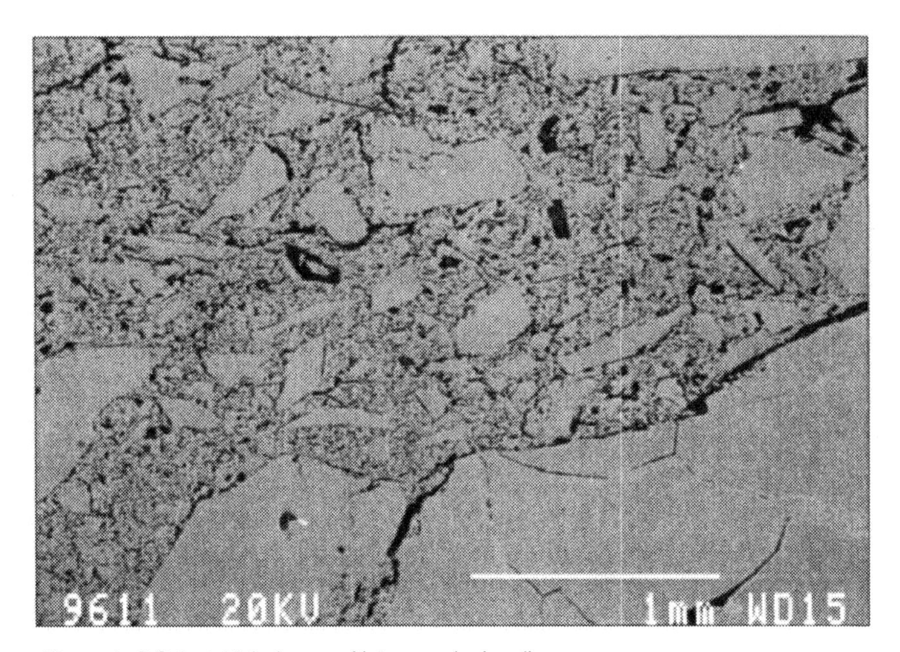

Figure 1. S Spinel, high degree of intergranular bonding.

Chemical Resistance

Spinel exhibits an inherent resistance to alkalis (Na,K +mixtures) which makes it an ideal product for the superstructure of oxy-fuel melters or indeed superstructure of conventionally fired melters where alkali is concentrated.

Silica is the traditional choice for melter superstructure; breastwalls, frontwalls, backwalls and crown. However in alkali laden environments particularly oxy-fuel; where an increase of volatile alkali species has been variously reported from 3 to 6 times, silica has exhibited accelerated wear.

This wear is created by concentrated alkaline hydroxides which react with the silica refractory to form sodium silicates. At melter temperatures in conjunction with high moisture content, these silicates exhibit a high degree of fluidity and therefore drip, eqn.1 below describes this process.

Spinel can react in a number of ways dependent on the accessibility and stability of the component oxides, however eqn.2 is postulated:-

$$SiO_2 + 2NaOH \rightarrow Na_2SiO_3 + H_2O \quad (1)$$

$$MgOAl_2O_3 + NaOH \rightarrow MgO + Na_2Al_2O_3 + H_2O \quad (2)$$

Equilibrium thermodynamic calculations were performed using the FACT package. The Gibbs energy minimization module EQLIBRIUM was used together with FACT databases.

Fig.2 and 3 consider 100g of silica or spinel solid phase respectively and examines the predicted phase constituents on successive additions of NaOH that is increasing the $NaOH/SiO_2$ or $NaOH/MgOAl_2O_3$ ratio, α, at a temperature of 1900K.

From the thermodynamic calculations; the fundamental intrinsic stability of spinel in alkali laden conditions versus silica is clearly shown.

In the case of silica only 1g of Na_2O is required (alpha θ.01) to digest 100 g, whereas for spinel requires nearly 36g.

To demonstrate this on a laboratory scale two vapour corrosion tests were conducted. In the first; finger test specimens were held over an alkali enriched glass, and in the

Second; block specimens were sealed over crucibles containing pure alkali (Na_2CO_3), and mixed alkalis (Na_2CO_3,K_2CO_3).

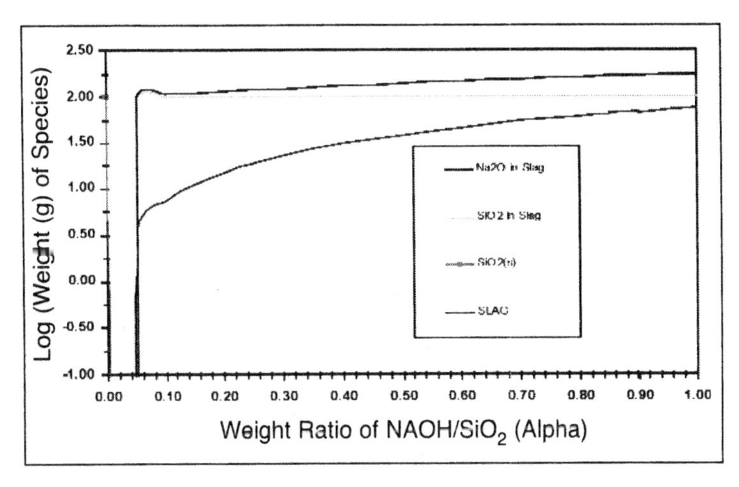

Figure 2. Predicted phase evolution of SiO_2-NaOH system at 1900K as a function of α.

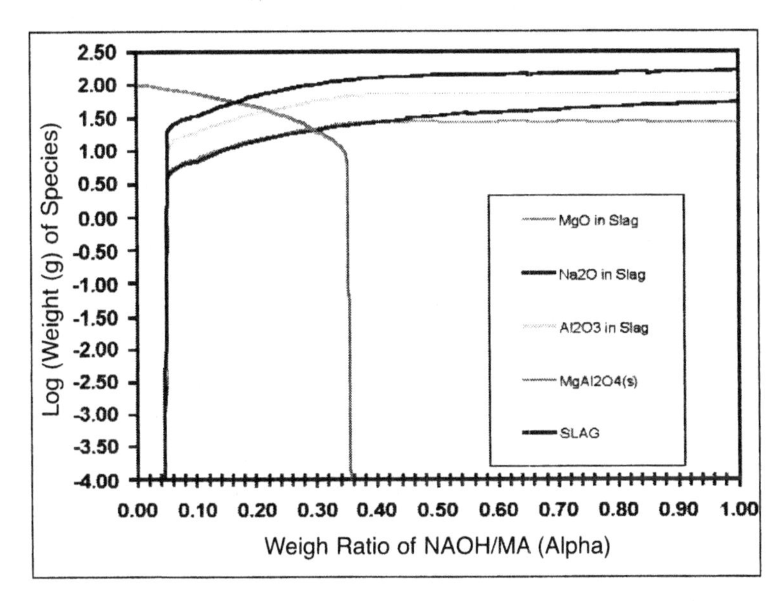

Figure 3. Predicted phase evolution of MA spinel –NaOH system at 1900K as a function of α.

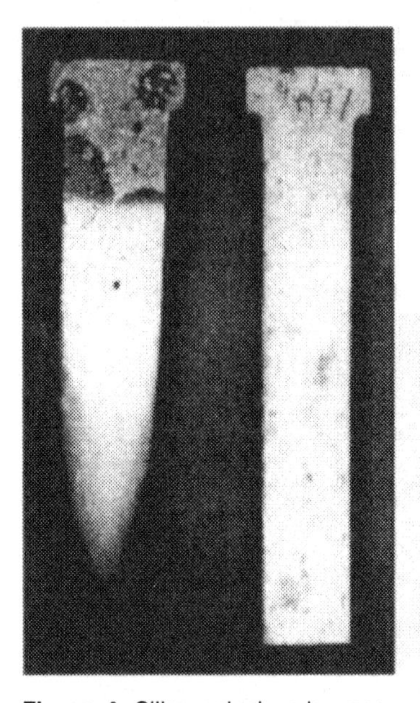

Figure 4. Silica spinel soda vapor test, 1500°C, 7 days.

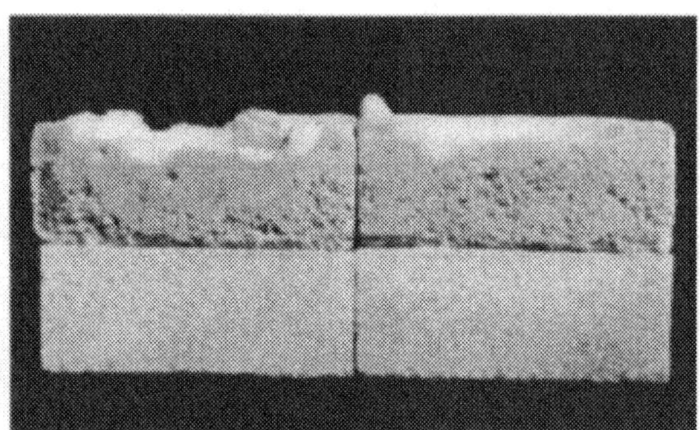

Figure 5. Vapor contact test pieces, 1600°C, Na_2O (left) Na_2O/K_2O.(right).

Fig. 4 shows finger test pieces that have been suspended over an enriched alkali glass (20%Na_2O), for a period of 7 days. Silica showed an average weight loss of 30%, the spinel composition displayed no weight change.

Fig. 5 shows the dissolution characteristics of silica (top) and spinel (bottom), against soda, and a soda/potash 50:50 mix at 1600°C. Silica is badly digested, whereas the spinel is unaffected.

The dissolution of silica in soda enriched environments is a well known phenomenon, however other volatile species also have considerable effects on the corrosion of silica superstructure in glass melters specifically boron oxide.

Boron oxide is not only an intrinsic ingredient in C and most E fibre glass compositions, but is also a minor addition in a wide variety of other glass compositions and can have a disproportionate effect on the corrosion of the silica superstructure.

The Equilib module and Fact databases of thermodynamic computational package FactSage™5.3 was used to predict the phase assemblages at 1600°C on successive additions of Na_2O and B_2O_3 (1/2 wt ratio) mixture.

Additions of flux to 100g of refractory from zero to maximum 30g were carried out in increments of 0.3 or 3.0g. The amounts of the predicted phases were plotted against the amount of flux. The products chosen for investigation were: Silica (SiO_2), Alumina (Al_2O_3), Mullite ($Al_6S_{i2}Ol_3$) and Spinel (Al_2MgO_4).

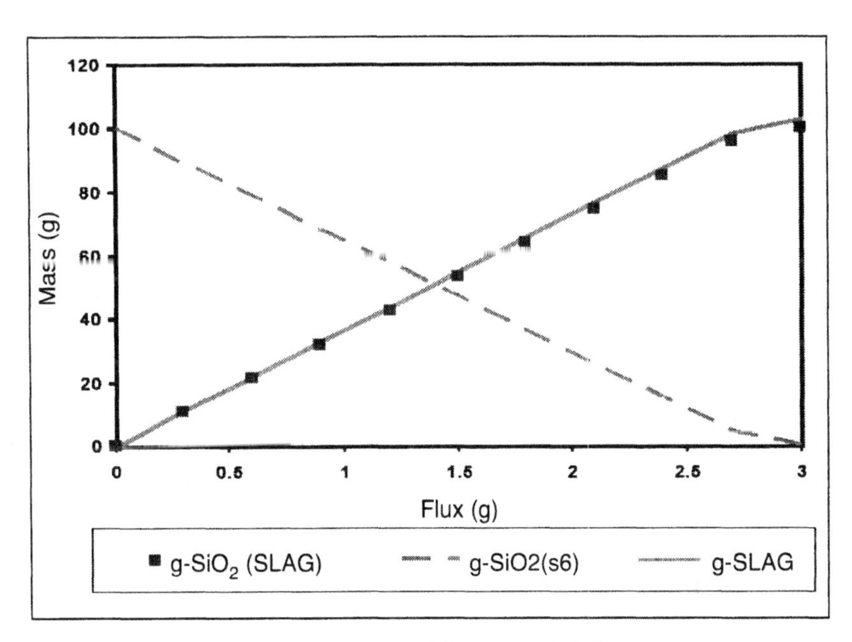

Figure 6. Liquid phase formation between SiO_2 and Na_2O/B_2O_3.

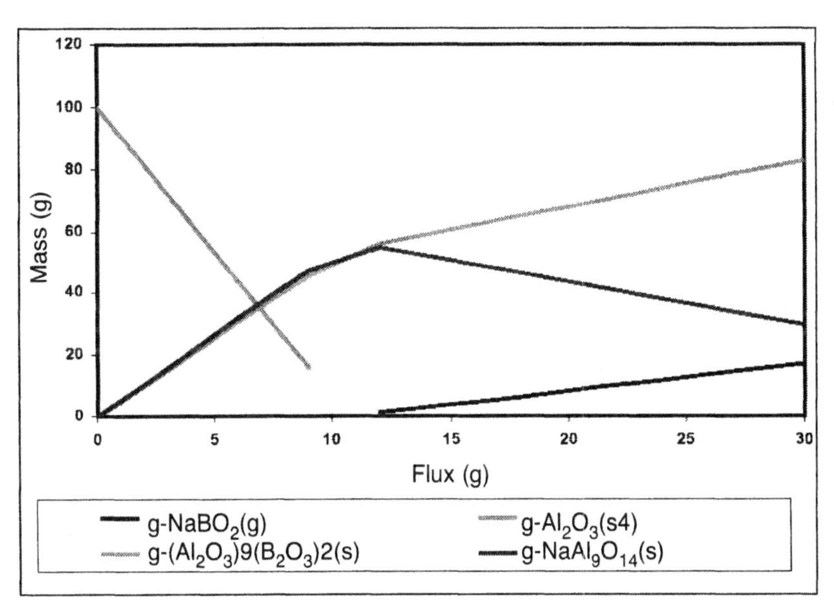

Figure 7. Liquid phase formation between Al_2O_3 and Na_2O/B_2O_3.

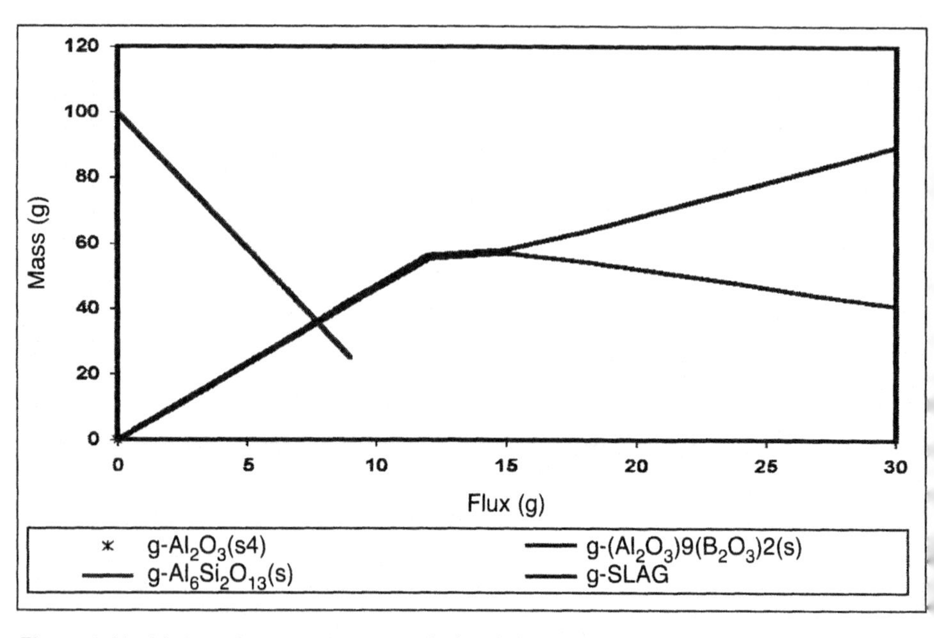

Figure 8. Liquid phase formation between $3Al_2O_3.2SiO_2$ and Na_2O/B_2O_3

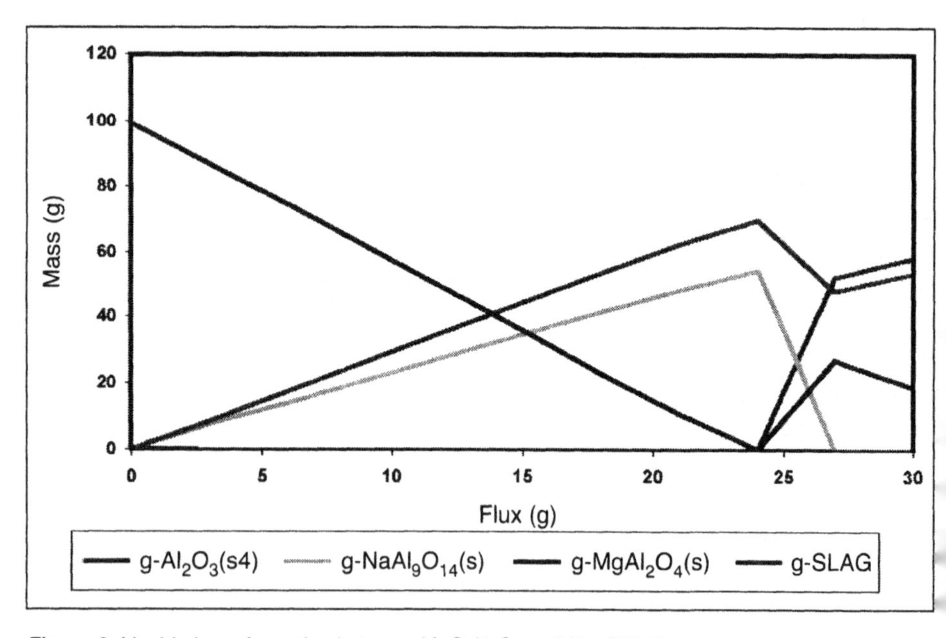

Figure 9. Liquid phase formation between $MgO.Al_2O_3$ and Na_2O/B_2O_3

From the plots Fig. 6 to Fig. 9, it is quite clear that from a thermodynamic point of view, spinel (Fig.9) is far more resistant to the Na_2O/B_2O_3 flux than the other refractory oxides in the study. Approximately 24 g of flux is required to digest 100 g of spinel, compared with 3 g for silica, 10 g for pure alumina and 12 g for mullite. Although these results should be used cautiously as they consider the equilibrium case and therefore no kinetic parameters, it forms a basis for further investigations.

A vapour corrosion test was performed to establish the relative resistance of silica and spinel to an aggressive combination of boron oxide and soda. For further comparison a fused mullite composition was included as this product type is commonly chosen for E-glass melter crowns.

Lids of nominal dimensions 55mm sq × 20mm deep were machined from the refractory samples. A groove was core drilled into one side of the "lid" to ensure a vapour tight fit onto a cylindrical crucible. 12.5g of a 30:70 mixture of B_2O_3 and Na_2CO_3 was placed into each crucible.

The assemblies (lids and crucibles) were heated at 200°C/hr to 1600°C and held for 24hrs. The appearance of the refractory lids is shown in Fig.10 below.

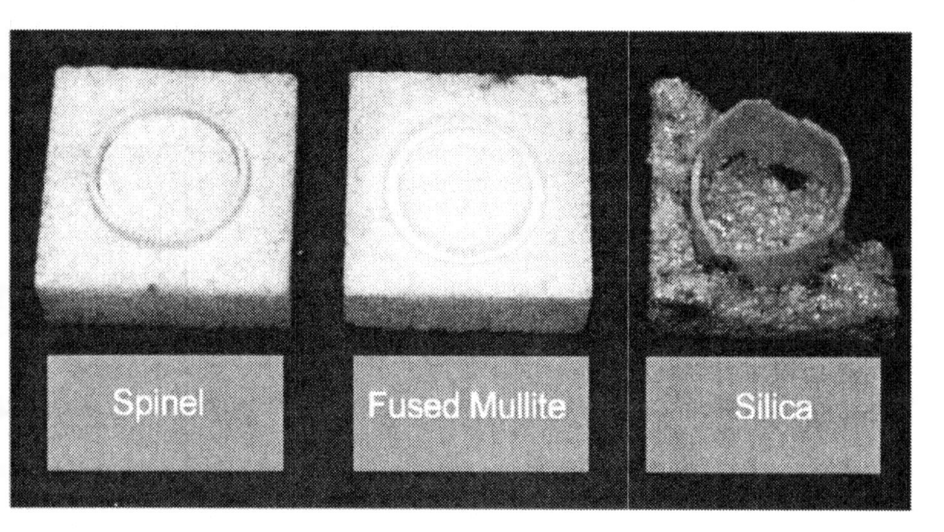

Figure 10.

The spinel sample showed no visible signs of attack;whereas the fused mullite surface was partially "etched" away although the sample had not swollen or indeed cracked. In comparison the silica sample had been digested (50%), leaving a "glassy" residue.

From the thermodynamic calculations undertaken together with the laboratory corrosion testing, spinel demonstrates significant potential for superstructure construction of melters producing a wide range of glass compositions.

Thermo-mechanical Properties

Spinel is a validated superstructure product with a current campaign of 6 years faultless operation. The next progression for spinel however is in the crown which is a developing role. A criterion for crown refractories is an inherent resistance to deformation under load at high temperatures. This property dictates the stability of the crown, and indicates the optimum level of crown insulation to ensure that the central load bearing third of the crown is maintained at a temperature of zero deformation.

Creep is the response of a refractory material to an applied stress, it is plotted in a number of ways, however the most descriptive is the percentage deformation over time as shown in Fig. 11.

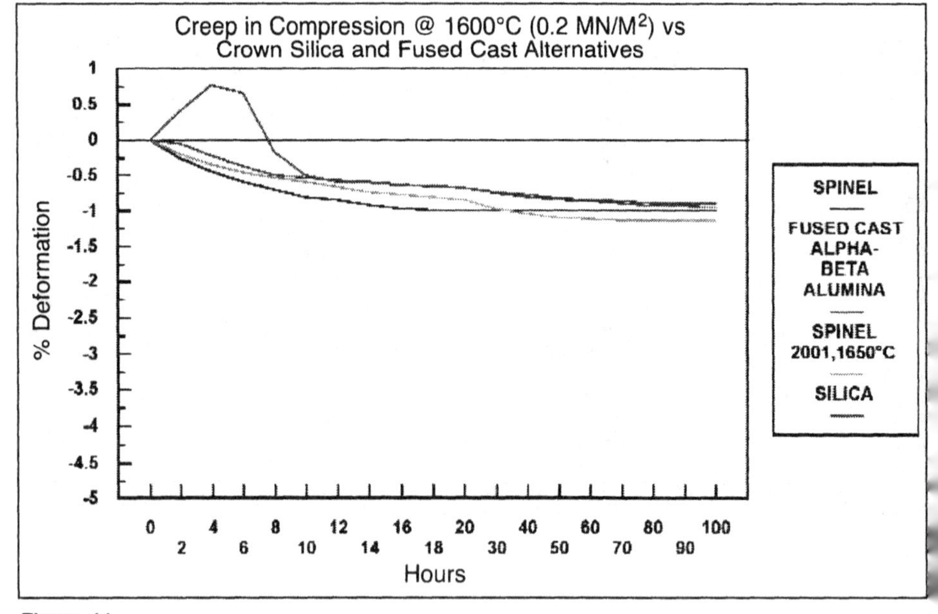

Figure 11.

The initial steep deformation (primary creep) is an elastic extension which reduces rapidly after a few hours. This is followed by a long period of steady state or secondary creep. It is the secondary creep which is critical to crown stability as this determines the long term potential for structural rearrangement of the refractory matrix. In this period of secondary creep as deformation can be so small it is often expressed as a creep number or indeed as a strain rate. Easy comparison is then made with current crown products (Table. 2).

In Fig. 11 spinel after the initial primary creep (18-20hrs) undergoes no further deformation up to 100hrs, and consequently both the creep number as defined and the strain rate over that time period are 0.

This is an excellent result in comparison with known crown products, which exhibit very minor deformation over the same time period.

Silica exhibits a strange initial expansion under load, and it has been postulated that this is a trydimite to cristobalite conversion.

TABLE 2

Product	Test Temperature (°C)	Creep Number	Log strain rate (/hr)
Spinel	1600	0	
Spinel	1650	0.8	-3.4
αβ Fused Cast	1600	1.6	-3.1
Silica	1600	2.2	-2.95

Note, creep number is defined as: $[(d_{100} - d_{50})/50] \times 1000$, where d_{100} and d_{50} are the deformations recorded at 100 and 50 hrs respectively.

Case Studies

Although spinel is established in the design of the melter superstructure for soda lime, it is also being trialled in the melter construction for a wide variety of other glass compositions.

To assess performance;panels are typically built into traditional superstructure with zircon splits, ram or mortar to separate dissimilar chemistries. Splits are recommended for horizontal loaded joints, whereas zircon mortar is quite acceptable for all vertical joints.

Melter Crowns

A melter crown in spinel is offered as a complete package of end arch, sealant layer, alkali resistant insulation and upper capping (typical construction is shown in Fig. 12).

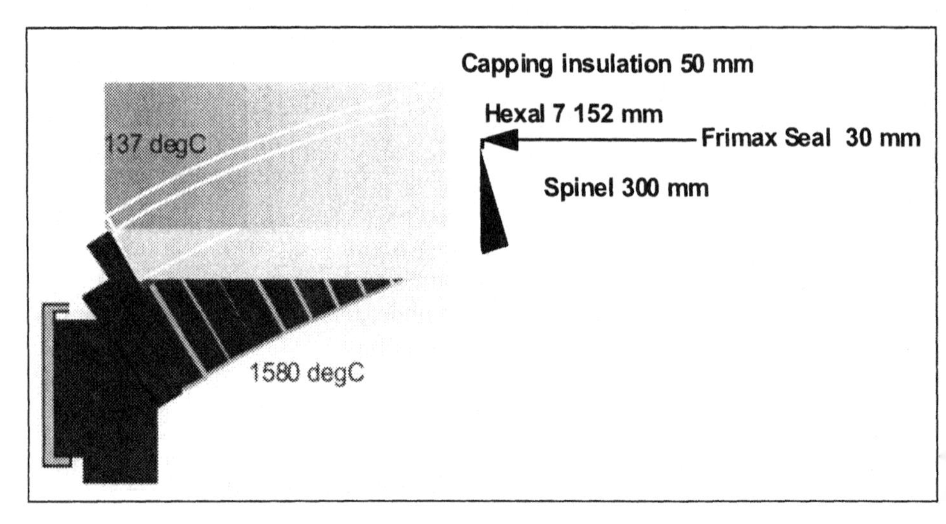

Figure 12.

The crown is installed as standard bonded end arch with edge pressed bricks to ensure accurate tapers. As the system is totally resistant to alkali there is no trepidation over preferential joint attack relating to mortar, damage to the front face of the brick or indeed condensation of volatile species in joint areas, and therefore altogether more tolerant to the vagaries of installation, including short time scales, distorted steelwork and unplanned events during heat-up.

The system is designed such that all potentially transmitted alkali will condense in the alkali resistant insulation. Crowns in spinel have now been in operation for 10 to 31 months without any reported problems.

Although spinel offers the ability to increase crown temperatures, and therefore load, the installations to date have been selected on the basis of chemical resistance alone as in the case study below:

- Crown: span 2.7 m, length 3.8 m operating at 1450°C
- Melter: end fired on natural gas with a single recuperator
- Glass:-soda-lime-silica with approximately 1% boron oxide
- Silica crown campaign 5 years, central zone replaced after 3.5 years and regular patching required.

Figure 13. Condition of silica at the end of the campaign illustrates the desire for change.

After 5 months of operation, the crown was inspected and there was no noticeable attack or run down on the hot face. At this stage silica crowns from previous campaigns would be showing signs of deterioration.

Burner Blocks

Spinel C was specifically developed for oxy-fuel burner systems in a pioneering forehearth designed to decrease pressure, increase heat input and reduce volatilization. Pilot scale tests determined that the traditional mullite bonded sillimanite burners were totally inappropriate with a life varying between hours and days.

Several other compositions were also trialled including cast AZS and cast Zircon. These products performed slightly better under oxy-fuel conditions, however due to the extreme temperatures the influence of eutectic melting compositions or decomposition brought eventual failure (cracks/deformation) of the burner.

Cast Spinel conversely proved to be stable on the test bed, and ultimately lasted 8 months in operation.

The initial product was cast using a hydraulic bonding medium and this found to promote grain growth coupled with deformation.

The refinement of this product was to convert to a no cement bond, and burners have operated for over two years with some replacement burners after the two year campaign.

Following the success of the forehearth burners;a natural progression was to explore the possibility of this product as a mid-campaign hot inserted burner for container glass melters.

After the removal of the original underport AZS burners, Spinel C replacements were set behind the final position to pre-heat. After 24 hrs the blocks were pushed into the firing position without any detriment to the integrity of the burner. The standard replacement burners based on AZS monolithic typically survived 6 months. Spinel C burners have given a minimum campaign of 2 years, and to date some blocks are operating after 3 years.

Fused Cast Repair

Spinel TSR is designed for hot repairs in fused cast superstructure. It has a thermal shock resistance of 9 cycles 950 C to water (S Spinel;5) which is comparable to fused mullite products commonly used in hot repairs and overcoats.

This product exhibits excellent alkali resistance (as S Spinel), and has been successfully hot inserted into an oxy fuel gas out-take;where fused cast AZS suffers significant wear due to alkali condensation and thermal shock (dampering due to boiler work).

Discussion and Conclusion

Spinel is an extremely versatile product with a wide range of applications in the glass industry. It has been utilised for melter construction on a global basis, specifically front and breastwalls for soda lime silica melters, with a validating campaign of over 6 years.

The advantages can be summarised as:-

- Inherently resistant to alkalis and boron compounds
- Mechanically stable at high temperatures, excellent creep resistance
- Standard construction, bonded end arch for crowns
- Low and linear thermal expansion
- Low defect potential;readily digested by silicate systems (on par with silica)
- Cost effective
- Extreme temperature stability, no eutectics, no dissociation

In addition;successful trials in borosilicate melters have determined the excellent stability of spinel which naturally has no silicate nature to be preferentially attacked. Complex shapes can be manufactured through a casting route, and thus burner, sight hole and camera blocks or indeed blocks outside of viable pressed dimensions can all be manufactured. Its use is not limited to the melter, and is relevant to all alkali condensation zones, or where thermal stability (creep resistance) is paramount.

In conclusion, the benefits of spinel in melter superstructure have been verified, however it has numerous potential applications and presents an exciting opportunity to resolve problems zones in melter design and repair.

BIBLIOGRAPHY

1. J.Boillet, W.Kobayashi, W.J.Snyder, C.A.Paskocimas, E.R.Leite, E.Longo, J.A.Varela, Ceram.Eng.Sci.Proc, 17[2]180-88 (1996)
2. G.Evans, Proc Int Congr. Glass Volume 1, 1-6 July 2001, 102-105
3. A.J.Faber, Ceram.Eng.Sci.Proc., 18[1](1997)

Getting Fired Up With Synthetic Silicates

John Hockman, Synsil Products, Inc., USA

Abstract

Alkaline earth silicates increase the flux:quartz ratio of any glass batch by replacing quartz. The flux:quartz ratio of a glass batch predicts the melting power or eutectic of a batch. An interesting technique that utilizes this effect allows the glassmaker to employ particularly hard to melt formulas without hindering melting in the furnace.

Introduction

Raw materials in the glass industry have changed little through the decades. Many aspects of glass manufacture are designed around the minerals that make-up a glass batch. Transportation, mixing, charging, firing, and other furnace operations are sensitive to the materials used.

Final glass chemistry is important to glass properties. The oxides of the glass control viscosity, liquidus, thermal expansion, index of refraction, and other key properties of the final product.

Chemistry changes are often dictated by final customer demand. Without changing raw materials, these chemistry changes often lead to a more difficult to melt glass batch. In the past, this was compensated through slower pull or higher temperatures. Higher temperatures, require more energy, and result in increased refractory wear and emissions. Quality and yield often suffer as a result of these moves. Changes to the glass batch often make the new furnace conditions more challenging.

Minerals like sodium carbonate and boric acid form a strong eutectic with silica to allow the glass batch to melt at a lower temperature. The ratio of sodium oxide or boron oxide to sand can give an indication how well at a given temperature that a batch will melt out.

The removal of CaO increases the one-on-one interaction of the sodium and silica. This increased interaction allows for a lower first melt temperature and overall faster melting in the batch.

Alkaline earth silicates as a replacement for dolomite or limestone can increase this flux:silica ratio by removing 10% to 50% of the free silica without changing final chemistry of the glass. This skewed eutectic ratio will increase the melt rate of the given batch.

The alkaline earth silicates most commonly used are diopside-type (to replace dolomite) and wollastonite-type (to replace limestone). Synthetic versions of both of these minerals also exist. Two of the more commonly used are SYNSIL® XD16™ composite mineral and SYNSIL® XW16™ composite mineral.

The SYNSIL® minerals as well as any synthetic or naturally occurring alkaline earth silicate can allow a glass chemistry change without compromising furnace operations or melt rate. These minerals give a "chemical boost" option to be weighed against the mechanical options and quality compromises that have been used to date.

Case Studies

In general, the concept of using a silicate to enhance melting, will increase yield or throughput while lowering energy demand- even when operating under no chemistry changes. These results have been documented in furnace trials in nearly all major glass segments. In this case, as described above, the increase in flux:silica ratio is used to leverage the glass formulator's ability to change the final chemistry without sacrificing furnace performance.

The two case studies that will be described below are modeled after a number of actual cases. Unfortunately, due to confidentiality agreements, the resultant furnace data cannot accompany this paper. By and large, furnace data confirmed the predictions of flux:quartz ratio.

Case 1: Reduced Alumina

The first formulation change deceptively changes the melt ratio of batch since it is not a direct change of the fluxing component, but results in the increase of silica by the removal of another mineral. In this case, the alumina was lowered to accomplish a glass property shift. Table 1 shows how the initial reduction in alumina requires the increase in silica that ultimately results in a 7.5% decrease in the flux:quartz ratio.

The decrease in sodium oxide to work on the silica resulted in a poorer melting batch. This new formulation required the need for significant furnace adjustments.

The third column of Table 1 indicates the advantage of replacing dolomite with a calcium magnesium silicate. While maintaining the new lower alumina glass, the batch using the SYNSIL® restored the flux:quartz ratio and improved it by 8.7% over the control and 16.2% over the new low alumina batch. This increase allowed for improved furnace operations. At the same glass temperatures and pull rate the defects (seeds and stones) were reduced. In another experiment, while maintaining quality and pull rate the glass temperature was decreased by 30°C.

Rather than have a poorly impacted furnace due to this chemistry change, the glassmaker realized an increased melt power and improved furnace operations with the alkaline earth silicate.

Table 1: Case 1 Decreased Alumina

*equal glass weights		Control	Decreased Alumina	Reduced Al; with *SYNSIL*® XD16
sand	(#)	1217.7	1316.0	1125.0
soda ash	(#)	452.7	449.5	454.9
SYNSIL® XD16	(#)	0.0	0.0	408.9
dolomite	(#)	329.3	328.5	0.0
sodium nitrate	(#)	18.7	20.2	18.3
kaolin clay	(#)	280.4	162.1	165.3
spodumene	(#)	45.0	48.6	44.0
sodium sulfate	(#)	11.1	12.0	10.9
fluorspar	(#)	13.2	14.2	12.9
borax (5 mole)	(#)	39.3	39.3	38.4
antimony trioxide	(#)	6.0	6.5	5.9
Fusion Loss	(%)	17.1	16.6	12.4
Flux:Quartz		0.235	0.217	0.256
Amount Quartz	(%)	50.5	54.9	49.2

7.5% Decrease From Control

8.7% Increase From Control

	Control	Decreased Alumina	Reduced Al; with *SYNSIL*® XD16
% SiO_2	68.42	70.74	70.62
% TiO_2	0.22	0.13	0.13
% Al_2O_3	5.95	3.75	3.78
% Fe_2O_3	0.12	0.08	0.09
% B_2O_3	0.96	0.96	0.94
% Sb_2O_3	0.30	0.33	0.30
% CaO	5.62	5.64	5.67
% MgO	3.59	3.58	3.65
% BaO	0.00	0.00	0.00
% Na_2O	14.34	14.30	14.37
% K_2O	0.04	0.02	0.02
% Li_2O	0.17	0.18	0.17

Case 2: Reduced Boron in E-glass

This case is more obvious in its impact on melting. For a variety of reasons, many continuous fiberglass producers have made the move to reduce boron. With every reduction in boron the glass batch becomes more difficult to melt: leading to either compromise in production and quality or increased furnace temperatures and wear.

Table 2 describes two steps of boron reduction in a generic E-glass formulation. In both steps, the B2O3:quartz ratio was severely reduced. If you have made the move to lower boron or are contemplating the move to boron reduction, the melt power and furnace operations in your plant are, or will be, altered.

Replacing the limestone with a silicate in the furnace will allow for an improvement (2.5 times) in melting from the low boron equivalent. In the 2.7% boron case, the furnace should see little ill effects from the boron reduction using the silicate since melt ratio was kept constant.

Table 2: Case 2 Reduced Boron E-Glass

	Control 6.3% B2O3	2.7% B2O3 No Synsil	2.7% B2O3 W/ Synsil	1% B2O3 No Synsil	1% B2O3 W/ Synsil
Material					
Sand	765	811	329	833	328
Limestone	765	852	0	894	0
Salt Cake	6	6	6	6	6
Clay	680	714	714	730	730
SYNSIL® XW16	0	0	1136	0	1192
Ulexite	340	147	147	54	54
	Wt %	Wt %	Wt %	Wt %	Wt %
SiO2	54.46	57.17	57.17	58.49	58.48
TiO2	0.53	0.55	0.56	0.56	0.56
Al2O3	13.19	13.75	13.76	14.02	14.04
B2O3	6.29	2.73	2.73	1.00	1.00
Fe2O3	0.33	0.32	0.32	0.32	0.32
CaO	23.08	24.24	24.23	24.80	24.79
MgO	0.47	0.37	0.37	0.31	0.31
K2O	0.19	0.14	0.14	0.11	0.11
Na2O	1.46	0.73	0.73	0.38	0.38
% Molar Alk	1.61%	0.84%	0.84%	0.46%	0.46%
% Batch Fusion Loss	21.74	20.97	14.27	20.59	13.46
B2O3 : Quartz	0.164	0.067	0.166	0.024	0.061
% Quartz reduction		-6.0%	57.0%	-9.0%	57.1%

Controls Melt Rate

2.5 X improved Melt Rate

Both of these examples can be generalized to other scenarios that could impact melt rate and furnace profit. The use of alkaline earth silicates can empower the glassmaker to obtain new glass formulas without compromising other furnace parameters.

Summary

Alkaline earth silicates unlock hard-to-melt glass formulations by increasing the eutectic between flux and quartz. Almost all changes to a formulation result in either sand being increased or the flux being decreased. Both changes result in a lower flux:quartz ratio and subsequent poorer melt performance. The melt power can be regained by substituting quartz with alkaline earth silicates.

Tall Crown Glass Furnace Technology For Oxy-Fuel Firing

H. Kobayashi, K. T. Wu, G. B. Tuson, and F. Dumoulin,
Praxair, Inc., Danbury, CT 06810, USA,
J. Böllert, Heye International GmbH, Germany

Praxair, in collaboration with Heye Glas, has developed an advanced oxy-fuel glass melting technology, termed "Tall Crown Furnace" (TCF) technology, to reduce silica refractory corrosion, NO_x emissions, SO_x emissions and particulate emissions while maintaining the high heat transfer and energy efficiency of oxy-fuel glass melting. The technology has been successfully implemented by Heye International into the engineering design of three commercial oxy-fuel fired container glass furnaces. The first furnace was lit off in early 1996 and currently has reached a campaign life of about 8 and a half years (over 9000 metric tons of glass pulled per square meter of the melter area) without major silica corrosion problems. The furnace is expected to operate for a total furnace campaign of 10 to 11 years with the original silica crown, with overcoat repair near the end of this period. The furnace has also demonstrated a high heat transfer rate and an excellent energy efficiency with a productivity as high as 3.5 metric tpd/m^2 (2.8 ft^2/tpd) without electric boost and a specific energy consumption of about 815 kcal/kg (3.3 MMBtu/ton) for flint glass with 60% cullet. The second and the third furnaces have adopted the same furnace design and have operated about six and four years respectively with furnace performance, similar to the first furnace.

Introduction

Although over 200 glass melting furnaces worldwide have been successfully converted from air-fuel firing to oxy-fuel firing since 1990,[1] accelerated silica crown refractory corrosion remains a significant concern for oxy-fuel firing. In some early oxy-fuel furnace conversions, major crown repairs were required within two years of a new furnace campaign. Often the most severe corrosion occurred at the joints between the crown bricks, creating so called "rat holes". The problem of localized joint corrosion has been mostly mitigated by better construction of the crown, proper heat up and better sealing after the furnace heat up. The primary cause for accelerated corrosion is believed to be the high concentration of alkali vapor species, especially NaOH and KOH, in oxy-fuel fired furnaces. In an oxy-fuel fired furnace, nitrogen in combustion air is largely eliminated and the volume of the combustion flue gas in the furnace is reduced to 1/3 to 1/5 of that of an equivalent air fired furnace. If the same amount of the alkali species volatilizes during glass melting, then the average concentrations of these species are increased several times in an oxy-fuel fired furnace. As a result, silica corrosion loss rates in a range of 3 to 25 mm per year silica were reported in a 350 tpd container glass furnace in the first oxy-fuel campaign.[2] In most furnaces, the highest corrosion rate was observed in the charge end, i.e., the coldest area of the furnace, while in some furnaces more corrosion was observed near the hot spot area.

In order to reduce the volatilization of alkali species in oxy-fuel fired glass melting furnaces Praxair initiated a series of studies at TNO Institute of Applied Physics, The Netherlands, to understand the mechanism of volatilization in glass melting furnaces in 1992.[3] The work was further expanded during 1995-1998 as Praxair organized an industry research consortium called "OXYGLAS" consisting of Praxair and several glass companies (AFG, Asahi, Borax, Corning, Manville, and Owens-Corning). The consortium funded a series of experimental laboratory studies and thermodynamic modeling studies to elucidate the corrosion mechanisms.[4-6] The volatilization and corrosion mechanisms developed from these studies were incorporated into Praxair's 3-D combustion space model of a glass furnace. The new model helped us to evaluate the potential furnace and burner design improvements which reduce volatilization and silica crown corrosion in oxy-fuel fired glass melting furnaces.[7] In later furnace conversions, several modifications were made to the furnace design and the oxy-fuel burners based on detailed CFD studies of alkali volatilization and actual furnace measurements of the concentration of alkali species.[8] The model was refined and applied in the design of a new 350 mtpd oxy-fuel furnace for Heye Glas in 1995 and the first Tall Crown Furnace (TCF) was jointly designed by Praxair and Heye International. To date the technology has been successfully implemented by Heye International into the engineering design of three commercial oxy-fuel fired container glass furnaces.[9] The first furnace was lit off in early 1996 and currently has reached a campaign life of about 8 and a half years (over 9000 tons of glass pulled per square meter of the melter area) without major silica corrosion problems and is expected to operate for a total furnace campaign of 10 to 11 years with the original silica crown of 375 mm thickness. The second and the third furnaces have adopted the same furnace design and have operated about six and four years respectively with good refractory conditions and furnace performance. A smaller TCF container glass furnace (150 mtpd) has been designed recently to rebuild an existing oxy-fuel furnace and is scheduled to start in 2005.

Key Design Features of Tall Crown Furnace Technology

The Tall Crown Furnace (TCF) Technology incorporates patented concepts to optimize the performance of oxy-fuel glass melting furnaces. The key design features are:

- High elevation of oxy-fuel burner placement above the glassmelt
- Tall crown height for more uniform temperature profile, i.e., to reduce the hot spot crown temperature and to increase the charge end crown temperature
- Partial furnace atmosphere stratification to create more oxidizing and lower H_2O vapor conditions near the glass-melt
- Praxair low momentum ultra low NOx JL Burners
- Single flue port to optimize energy efficiency
- Optional Side-wall Cooling Air Deflector ("SCAD") to reduce air infiltration into the furnace for NO_x reduction
- Optional air or oxygen bubblers

Since the accelerated corrosion of silica refractory in oxy-fuel fired furnaces is caused by the fluxing of silica bricks and joints by alkali vapors in the furnace atmosphere, the first design goal is to minimize the source, i.e., volatilization of alkali species from the glass melt. The second design goal is to minimize the transport of alkali vapors to the silica crown areas. The third design goal is to reduce the mass transfer rate and reactions of alkali vapors with silica bricks. Praxair TCF technology incorporates several important design features to accomplish these goals. Since the TCF reduces the alkali volatilization by about 50% as compared with the conventional oxy-fuel furnace, particulate emissions are also reduced by about 50%. Overall radiative heat transfer is improved by the TCF design, resulting in high energy efficiency in spite of a small increase in furnace wall heat losses. The NO_x emissions are reduced to 1/3 to 1/4 of those from the conventional oxy-fuel burners with the Praxair JL Burners. In the following sections, engineering principles used for the TCF design and actual results from commercial TCF furnaces are described.

Reduction of Alkali Volatilization and Particulate Emissions

Alkali vapors (mostly NaOH and KOH) in the soda lime glass furnace are mainly formed by the reactions of water vapor with alkali species (e.g., Na_2CO_3, Na_2SO_4, Na_2O, K_2O, etc.) in the batch and glassmelt. As the flue gas from a glass furnace cools down in the regenerators and flue ducts, NaOH and KOH vapors react with SO_2 and O_2 in the flue gas to form Na_2SO_4 and K_2SO_4, which subsequently condense to form sub-micron size particles in the stack. Typically 80 to 90% of the particulate emissions are alkali sulfate particles.[3] Thus, by reducing alkali volatilization both particulate emissions and silica crown corrosion rates can be reduced. The volatilization rate of NaOH is believed to be controlled by the mass transfer rate of NaOH vapor in the boundary layer near the glassmelt surface to the bulk furnace atmosphere. For a fixed glass composition, three key parameters controlling the alkali volatilization rate are (1) temperature of the glassmelt surface, (2) velocity of furnace gas near the glassmelt, and (3) water vapor concentration in the furnace atmosphere near the glassmelt surface. All of these three parameters are reduced by the TCF design.

In Figure 1, measured particulate emissions from regenerative air fired furnaces, conventional oxy-fuel fired furnaces and advanced oxy-fuel fired furnaces are shown as a function of the specific glass pull rate. It is well known that particulate emissions increase sharply with the specific glass pull rate as a result of increased glassmelt surface temperatures and higher gas velocity under higher pull rates. A comparison of the data from air fired furnaces and those from conventional oxygen fired furnaces shows about 20 to 30% reduction of overall particulate emissions under oxygen firing.

The reduction is mainly attributable to the lower gas velocities in oxy-fuel fired furnaces. In these early oxy-fuel conversions, little design consideration was given to minimizing the alkali volatilization and particulate emissions. In later furnace conversions, several modifications were made to the furnace design and the oxy-fuel burners. The design improvements included the key features of the TCF design, i.e., lower flame velocities, a higher burner elevation and a taller crown height. Some of the improved results are also plotted along the line of "Oxy-Furnace New Design" in Figure 1. Particulate emissions were reduced approximately by a factor of two compared to the earlier oxy-fuel furnace design. At a specific pull rate of 3mton/day/m^2 particulate emissions of 0.25 kg/mton (0.5 lb/ton) were achieved. Further reductions to 0.1 to 0.15 kg/mton (0.2 to 0.3 lb/ton) seem feasible with other design improvements, especially when batch/cullet preheating is integrated to reduce the firing rate and the glassmelt surface temperature.

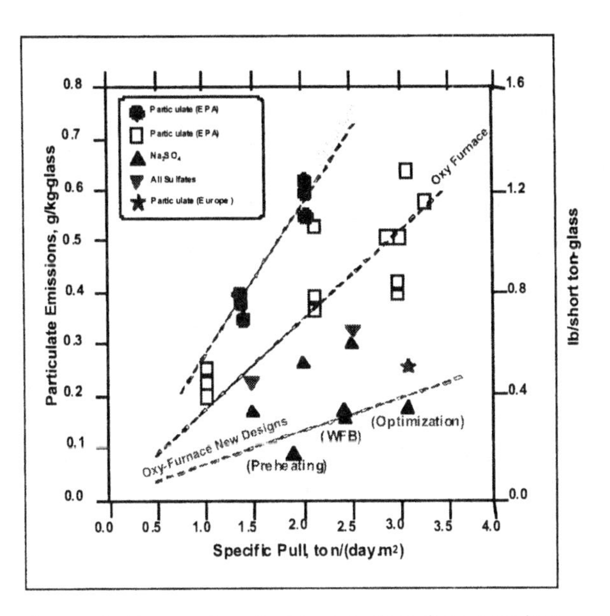

Figure 1. Measured particulate emissions from container glass furnaces.

Effects of Burner Elevation

The effects of burner types and furnace geometry were studied by an alkali volatilization and corrosion model.[8] Figure 2 shows a comparison of volatilization rate at the glassmelt surface for three different burner elevations. The model predicted the areas of high alkali volatilization rates underneath each flame in an oxy-fuel fired container glass furnace, caused by higher glass surface temperature and higher convective velocity. As the burner elevation was increased, the "hot spots" on the glassmelt surface created by the oxy-fuel flames and the gas velocity near the glassmelt surface were reduced. Although the average glassmelt surface temperature remained approximately constant for a given pull rate, the overall alkali volatilization rate was reduced by about 50% by raising the burner height above the glassmelt.

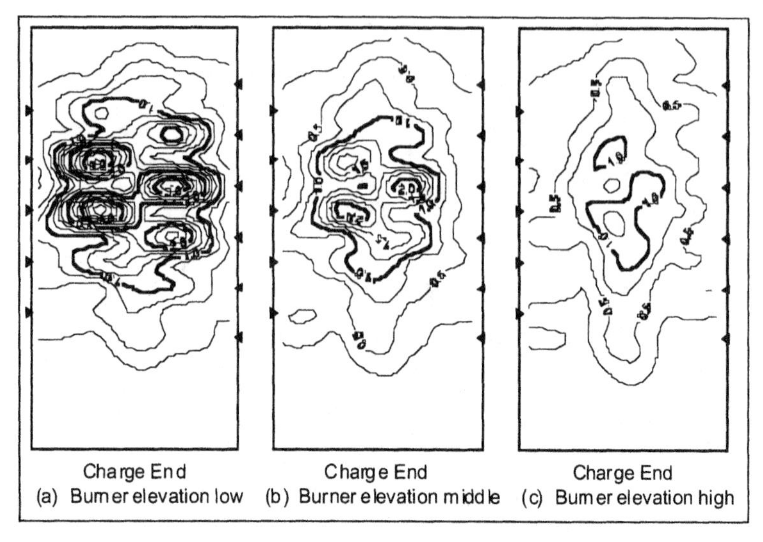

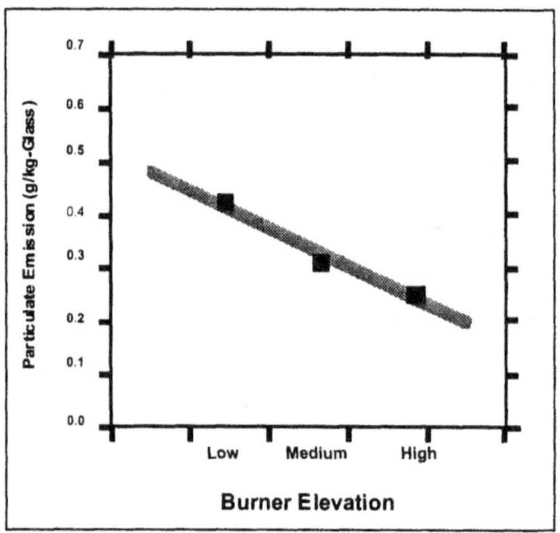

Figure 2. Calculated Rate of NaOH volatilization and particulate emissions versus burner elevation.

Furnace Atmosphere Stratification to Reduce H_2O Concentration

The average concentration of water vapor in an air fired glass melting furnace is about 15 to 20% and that in an oxygen fired furnace is about 50 to 60%. The high water vapor concentration increases alkali volatilization, as the mass transfer controlled rate of volatilization is approximately proportional to the square root of the partial pressure of water vapor. Praxair developed a burner technology using deeply staged combustion in which most of the combustion oxygen is injected separately from the main burner toward the glassmelt surface to achieve ultra low NO_x emissions. This patented process[10-11] creates a partially stratified oxygen rich atmosphere near the glassmelt surface and reduces the water vapor concentration near the glassmelt surface, and thus helps to reduce the rate of alkali volatilization.

Silica Crown Refractory Corrosion

Corrosion of silica crown refractory under oxy-fuel firing is caused by the fluxing of silica bricks and joints by alkali vapors in the furnace atmosphere. The OXYGLAS consortium studies at TNO[4-6] showed:

- The corrosion rates of silica bricks and joints are strongly related to the mass transfer rate of alkali vapor species to the bricks.
- The calcium rich binding phase (beta-wollastonite, $CaO.SiO_2$) used in the regular silica brick is preferentially attacked by alkali vapor species, forming a glassy phase which penetrates through the silica grain boundaries and dissolves silica grains.
- The rate of silica brick loss appears to be controlled by "washing" of the glassy phase. At high temperatures a relatively small amount of alkali oxides in the glassy phase reduces the viscosity of slag sufficiently to allow the slag to flow. Thus, a small amount of alkali vapor transferred to the brick surface can "wash" a large amount of silica.
- In the colder area of the crown near the charge end below about 1475°C, a glassy layer containing a high alkali oxides concentration (up to 14%) forms with a low enough viscosity to flow. Thus, the amount of silica loss per unit amount of alkali vapor transferred is reduced substantially. However, the alkali vapor concentration and the rate of gas phase mass transfer are often higher in the charge end of the furnace. Thus, the mass transfer rate of alkali vapor to the brick, and the viscosity and temperature of the glassy phase control the rate of silica loss.
- Fused silica reacts less intensively with alkali vapor species. The initial rate of alkali vapor absorption was about 1/4 of that for the regular silica brick. At temperatures below about 1450°C sodium silicate will be formed. At higher temperatures less alkali vapor attack is expected since sodium silicate is thermodynamically unstable. Fused silica bricks or special silica bricks with very low calcium content are preferred over the regular silica brick for the crown.

Recent studies at Sandia National Laboratories[12] generally support these findings and the rate controlling step for silica crown corrosion is believed to be the mass transfer rate of alkali vapors (mostly NaOH) in the boundary layer near the crown surface. This process is analogous to the mass transfer controlled volatilization of alkali species at the glassmelt surface. In order to reduce the mass transfer rate of NaOH from the furnace atmosphere to silica crown bricks, three key measures are necessary: (1) to increase the crown surface temperature, but don't exceed the temperature of glassy phase flow (2) to reduce the concentration of alkali vapors near the crown, and (3) to reduce the velocity of gas near the crown. A higher crown temperature increases the equilibrium vapor pressures of NaOH at the glassy phase formed in the silica brick and reduces the concentration gradient of NaOH between the bulk furnace and the crown surface. Thus, the mass transfer rate is reduced, as long as the glassy phase is stable and does not flow. If the glassy phase becomes fluid enough to wash out, a fresh silica brick surface is exposed and the equilibrium NaOH concentration is reduced and a high gas phase mass transfer rate of NaOH and a high rate of silica loss would result. There is a maximum temperature limit as a small amount of NaOH transferred to silica bricks "washes" a large amount of silica at high temperatures.

Figures 3 and 4 show silica brick loss measured by post-mortem analyses of crown bricks at different furnace lengths on a 350 tpd oxy-fuel fired furnace operated by Gallo Glass between 1991 and 1997.[2] The corrosion rate decreased from 33 mm/yr in the charge end to about 4 mm/yr near the hot spot as the estimated crown temperature increased from 2450°F to 2860°F. There was an indication of minimum corrosion in a temperature range of 2820°F to 2860°F in this furnace and the corrosion rate at 2880°F was about 13 mm/yr. Since the NaOH concentration and the gas velocities in the furnace varied in different areas within the furnace, the large changes in the corrosion rate along the furnace length is not necessarily caused by the crown temperature alone. However, the strong correlation with temperature suggests that the silica corrosion rate could be reduced several fold by maintaining the optimum temperature range for the entire crown.

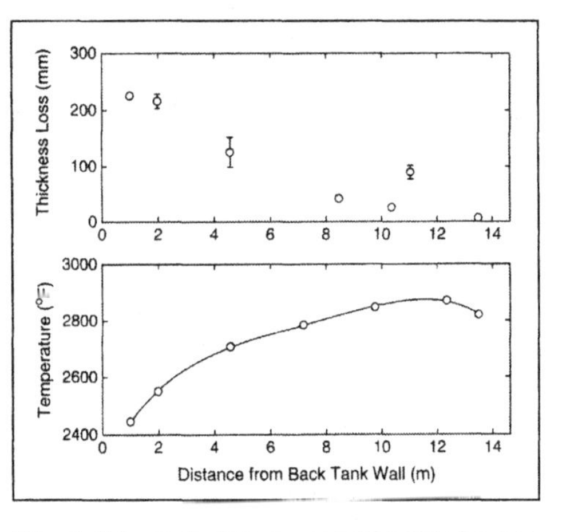

Figure 3. Loss of silica crown bricks after furnace campaign.

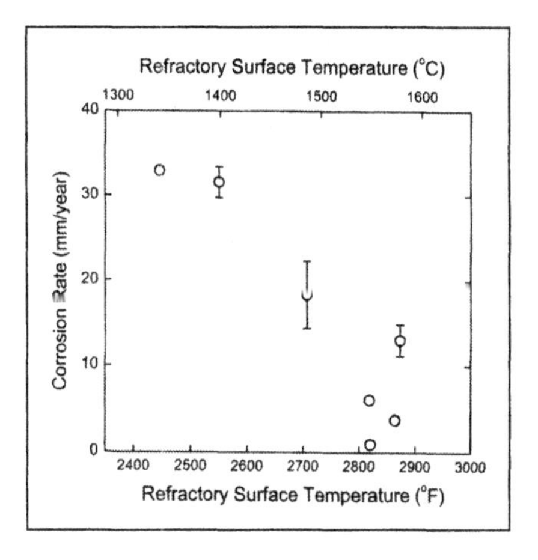

Figure 4. Corrosion rate versus refractory surface temperature.[2]

Tall Crown (Breastwall Height)

The crown height is an important design parameter for silica crown corrosion as well as furnace heat transfer. In the furnace design study using a 3-D combustion space model, the alkali volatilization rate was reduced by about 50% by increasing the burner height. However, with constant crown height the crown temperature increased as the burner height was raised, approaching the maximum refractory temperature limit of about 1600°C at the hot spot of the furnace. It was found that raising both the burner and the crown heights offered several advantages in reducing the crown corrosion rate while maintaining good heat transfer characteristics.[13] In general a higher crown height results in a lower average crown refractory temperature and a flatter temperature profile along the furnace length, i.e., the charge end temperature is increased and the hot spot temperature is reduced. Since the maximum glass pull rate is often limited by the peak crown temperature, a lower crown temperature enables a higher firing rate and an increased glass pull, which is an additional benefit of the TCF design.

In addition to the improved temperature profile, reduction of both the NaOH concentration and the convective velocity near the crown are other important features of the TCF design to reduce the mass transfer rate of NaOH to the silica crown. For silica crown corrosion the important parameter is the concentration of the alkali vapor near the crown, not the average concentration in the flue gas. There are large differences in the NaOH profile between air and oxy-fuel firing. In a cross-fired regenerative furnace, each flame from a side wall port has a short straight path to the opposite sidewall. The NaOH vapor from the glassmelt or batch is concentrated near the melt surface and is exhausted quickly from the opposite port. As a result, the NaOH concentration near the crown is much lower than the average furnace NaOH concentration. In a conventional oxy-fuel fired furnace, the flue gas is exhausted from one or two flue ports located near the charge end. NaOH vaporized in the hot end of the furnace is mixed with several oxy-fuel flames as the bulk furnace gas flows toward the flue ports. As a result, the furnace atmosphere is relatively well mixed, resulting in relatively high concentrations of NaOH near the crown refractory. In a mathematical model study, the NaOH concentrations near the crown refractory of a conventional oxy-fuel fired furnace was shown to be an order of magnitude higher than those of a corresponding cross-fired regenerative air furnace, when the average NaOH concentration in the oxy-fuel fired furnace was three times higher.[7]

In the TCF design, the low flame momentum of the oxy-fuel burner and the tall crown reduce the circulation of NaOH and other volatile species from the glassmelt surface to the crown and the NaOH concentration and the gas velocities near the crown are reduced. Both of these factors contributed to the reduction of the mass transfer rate of NaOH to silica crown and the corrosion rate.

Heat Transfer and Energy Consumption

Traditionally long and wide luminous flames covering the entire glassmelt surface area are believed to be more efficient and are preferred for glass melting furnaces. To emulate this design philosophy, oxy-fuel flames with high luminosity and wide flat flame shapes have been developed and installed in many furnaces.[14] The high burner elevation in the TCF design raised concerns about the possible loss of heat transfer to the glassmelt. Contrary to the conventional wisdom, good heat transfer efficiency is maintained with the TCF design. Although the larger distance between the flame and glassmelt surface reduces convective heat transfer and direct radiative heat transfer from flame to glassmelt, bulk gas to glassmelt radiative heat transfer increases because of increased gas emissivity due to the longer "beam length" afforded by the tall crown height. The net effect is little change in the overall heat transfer efficiency in the flame regions. In the charge end of the furnace near the flue gas exit port, the heat transfer efficiency by bulk gas to the cold charge is increased due to the higher gas emissivity and longer gas residence time caused by the larger furnace volume. With proper flue port placement, the net effect is a reduction in flue gas temperature despite higher charge end crown temperature. The decrease in flue gas temperature corresponds to an increase in available heat, which offsets a small increase in the furnace wall losses caused by the increase height of the breast walls which are generally very well insulated. More detailed discussions on heat transfer issues can be found elsewhere.[1,15] Table 1 shows a comparison of the furnace energy balances of a 330 tpd (300 mtpd) conventional oxy-fuel fired and Tall Crown container glass Furnace.

TABLE 1. ENERGY BALANCE COMPARISON BETWEEN CONVENTIONAL AND TALL CROWN CONTAINER GLASS FURNACE

	Conventional Oxy-gas	Tall Crown Design
Glass Pull Rate (tpd)	330	330
Cullet ratio (weight % of charge)	30	30
Flue gas temperature (F)	2620	2600
Fuel Input (MMbtuh)	53.14	53.28
Energy Outputs (MMbtuh)		
Energy to batch reactions and glass	24.64	24.64
Tank & Furnace structural heat loss	6.60	6.84
Flue gas sensible and latent heat	21.90	21.80
Total	53.14	53.28
Specific Energy (MMbtu/ton)	3.86	3.87

NO_x Emissions

The Praxair JL Burner provides a unique design for very low NO_x emissions and partial furnace atmosphere stratification. The burner design utilizes the patented concept of "deep oxygen staging"[16] and a separate injection of secondary oxygen under the low temperature rich primary flame. The rich primary zone stoichiometry produces a low momentum high luminosity flame for high heat transfer efficiency and the separate secondary oxygen injection produces a long flame with a higher oxygen concentration and a lower water vapor concentration near the glassmelt surface. NO_x emissions from JL Burners are approximately 1/3 to 1/10 of those from the conventional oxy-fuel burners.

Heye Glas's Experience with Tall Crown Container Glass furnaces

Based on model predictions and other engineering considerations, a new 350 mtpd furnace with a tall crown for melting flint container glass was built by Heye Glas, Obernkirchen, Germany in 1996.[9] Measurements of alkali vapor species and particulate emissions confirmed a substantial reduction in the alkali vapor concentration. After more than eight years of operation the silica crown in this furnace is still in good condition and is expected to last over 10 years. The original thickness of the silica crown bricks was 375 mm. Heye commissioned a second tall crown oxy-fuel furnace for green glass in 1998 in the Netherlands. A third tall crown oxy-fuel furnace for melting amber glass was commissioned in April 2000 in Obernkirchen.

Heye had specific reasons for its decision to adopt oxy-fuel technology. One reason was to avoid having to install secondary deNO_x measures which would otherwise be required to satisfy the regulatory requirement for NO_x emissions below 500 mg/m? flue gas for new installations. The second reason was the possible reduction of melting costs and the necessity to place a furnace with high melting capacity into an existing building with a very limited space.

The preconditions within Heye for engineering the new furnace were quite good given Heye's experience in design and operation of continuously heated recuperative furnaces with adapted heat recovery via steam generation. Heye viewed steam generation heat recovery technology as transferable to oxy-fuel furnaces, and in fact waste heat boilers have been integrated with Heye's oxy-fuel furnaces. Of course the tasks for implementing the new oxy-fuel process were monitored very intensively and with a certain respect, and design parameters and technical solutions were carefully selected.

To prevent faults in the design of the new furnace, Heye carried out tests of several market-proven oxy-gas burners in an existing air-fuel furnace and inspected wall temperatures, crown temperatures near the burner, and flame development with an ultraviolet camera. The primary purpose of this test was to prevent overheating of gable walls and crown by excessively high temperature and poorly distributed flames. Heye selected Praxair's JL-Burner based on this investigation.

Furnace modeling based on Heye's design parameters was carried out by Praxair to justify the selected furnace dimensions, burner position and flue location. Heye's experience in furnace design led to the assumption that a very important question during the modeling was the dimensioning of the combustion chamber. Avoiding temperatures below 1430°C was considered a main concern of the design. To prevent the formation of tridymite which builds up in connection with the availability of sodium and water, available as NaOH, in a much higher concentration than in the nitrogen-diluted flue gas of a conventional air-fuel furnace. This avoidance of regions of low temperature was identified as the most important criteria to ensure the reliability of the silica crown.

High attention was also given to the reliability of the flue gas systems. Experience with increases in oxygen concentration due to furnaces using high sulphur-fuels, having similar SO_x concentration as oxy-gas furnaces, showed that infiltration of fresh air is responsible for most of the disturbances appearing in the flue gas systems behind the furnaces such as clogging or corrosion. The optimum system to avoid those problems seemed to be the heat transfer in a steam boiler in which all sensible temperature ranges between 1400°C and 200°C could be passed without any addition of fresh air. Long time experience with similar systems was used to design a new waste-heat boiler with automatic cleaning device together with a partner.

To achieve optimal results regarding energy consumption (not only gas but in this case also oxygen and respectively, electricity), NO_x emissions and melting performance, many details in the selection of the techniques and components had to be considered. All openings of the furnace are kept sealed by means of newly designed batch chargers and peep-hole flaps. Independent burner controls were built which use an independent ratio controller for each burner operating very precise motor-operated control valves.

Experiences and Operational Data

The furnace melting flint glass feeds two IS 20-section DG forming machines producing 1l and 0.33l mineral-water bottles. From the start-up on, the furnace has melted the desired amount of flint glass with excellent glass quality. Seed counts are on average below 20/oz. The two forming machines pull 330 metric tons per 24 hours on average. The furnace is operated with approximately 60% cullet powder of external cullet from Heye's cullet pulverizing plant and there have been almost no defects due to inclusions in the glass nor any customer claims over the whole period of the operation.

The change from conventional to oxy-fuel combustion has not resulted in significant changes in the physical properties of the glass. The influence of the higher water content in the furnace atmosphere has been monitored carefully, but the impact on workability or any reduced tendency to specific defects could not be proved. The amount of sulfate for refining purposes was reduced during the commissioning relative to the amount used in air-fuel melting.

Energy consumption figures are even better than expected. With the parameters mentioned above, the average consumption is around 3.35 mJ/kg of glass (800 kcal/kg) which results in melting costs lower than those all other conventional furnaces in operation. This capacity and consumption is achieved at a pull rate of above 3 tonne/m^3 and without any electric boosting.

Concerning emissions data, especially NO_x, all expectations have been more than fulfilled. While the authorities agreed upon a regulation in which the legally demanded target of 500 mg NO_x/m^3 of flue gas for conventional furnaces was converted to 0.7 kg NO_x/ton of glass based on same mass-flows of NO_x for both furnace types, the results achieved surely meet these limits. Measurements show emission figures between 0.5 kg NO_x/ton and 0.35 kg NO_x/ton of glass using natural gas containing about 11% N_2.

For all other emission components, especially those arising from the melt such as particulates, SO_x, HCl and HF, no major changes were expected. In fact, the mass-flow of those emissions is almost equivalent to the emission as it would have been measured on a conventional air fired furnace with the same operation parameters. This high density of emissions leads to particular operating conditions as expected:

- The corrosiveness of flue-gases to the crown material is as strong as expected and the influence of the temperature is as well. The calculations of the modelling have proven to be correct and all risks are under control.
- The fouling of the flue-gas systems is as expected. The reserve kept in the design of the boiler heating surfaces is sufficient and the automatic cleaning system works satisfactorily. Manual cleaning requirements can be kept within limits.
- The bag-filter behind the system is able to keep all emissions within the permitted limits.

Waste-Heat Recovery Boiler and Power Generation With a Steam Turbine

The operation of the steam-boiler is quite smooth. The steam production is typically 5.5–5.8 tons/hour and is therefore not so much different from the steam production of a comparable conventional furnace. The boiler operation has proven to be very reliable. The furnace has to work together with the waste heat boiler based on a safety concept which has been agreed to by the authorities. The boiler produces steam at a pressure of 30 bar and a temperature of 430°C. All sensitive shut-off functions are transferred into the monitoring and operational system so that the condition of the equipment is always visible. Based on these functions, an automatic start of the furnace considering all necessary safety interlocks has been installed.

The steam from the boiler is expanded through a steam turbine to generate electric power. About 990 to 1040 kW of power is generated from the steam turbine which supplies the power required for the on-site cryogenic oxygen plant and other plant lighting and equipment. The furnace and the control equipment have been accepted quite well by the plant people.

Performance Goals for the Next Generation Tall Crown Container Glass Furnaces

In Table 2, projected characteristics of an optimized TCF oxy-fuel fired container glass furnaces are shown. They were extrapolated from actual measurements taken from advanced oxy-fuel fired container furnaces with the effects of a batch/cullet preheater/filter system[14]

TABLE 2. PROJECTED PERFORMANCE OF OPTIMIZED TCF CONTAINER GLASS FURNACE

Furnace Capacity	350 mtpd (385 tpd) @ 60% cullet flint glass
Productivity	4.0 mtpd/m^2 (2.5 sq ft/tpd)
Energy Efficiency	625 kcal/kg (2.5 MMBtu/ton)
NO_x Emission	0.05 g/kg (0.1 lb/ton)
SO_2 Emissions	0.4 g/kg (0.8 lb/ton)
Particulate Emissions	0.1 g/kg (0.2 lb/ton)
Furnace/Refractory Life	10 to 15 years (equal to air furnace)

Most of the improvements over the state-of-the-art oxy-fuel furnaces are direct or indirect benefits of batch/cullet preheating. Up to 30% of fuel and oxygen could be saved by fully recovering waste heat from a directly fired oxy-fuel furnace. Since batch/cullet preheating reduces the heat transfer requirement in the furnace, a higher specific production rate can be achieved, as demonstrated in air fired furnaces quipped with batch/cullet preheater. Similar to electric boosting, batch/cullet preheating reduces the glass surface temperature at a constant production rate. The reduction in the firing rate proportionally reduces the combustion gas volume and reduces the average gas velocity in the furnace. Lower glass surface temperature and lower gas velocity both help to reduce the volatilization of alkali species from glassmelt and batch. Thus, particulate emissions are reduced and the potential for silica crown corrosion is also reduced. The net economic benefit from preheating, of course, depends on the capital investment required for the preheating equipment and its installation costs. An economic study15 showed that the total capital cost of a new oxy-fuel fired furnace equipped with a batch/cullet preheater/filter system is still significantly lower than the capital cost of the traditional regenerative air fired furnace.

Summary

The performance of oxy-fuel fired glass melting furnaces has improved significantly with the Tall Crown Furnace design. The improved burner/furnaces designs reduced corrosion of silica crown and the life of a well designed oxy-fuel fired container glass furnace is now expected to be over 10 years. Praxair JL Burners reduced NO_x emissions substantially, improved heat transfer characteristics and reduced volatilization of alkali vapors from glassmelt and batch. Particulates and SO_2 emission were reduced over 50% as a result of reduced volatilization of alkali species and a reduced sulfate requirement for fining. The technology has been successfully implemented by Heye International into the engineering design of three commercial oxy-fuel fired container glass furnaces. The first furnace was lit off in early 1996 and currently has reached a campaign life of about 8 and a half years (over 9000 tons of glass pulled per square meter of the melter area) without major silica corrosion problems and is expected to operate for a total furnace campaign of 10 to 11 years with the original silica crown. The furnace has also demonstrated a high heat transfer rate and an excellent energy efficiency with a productivity of as high as 3.5 metric tpd/m^2 (2.8 ft^2/tpd) without electric boost and a specific energy consumption of about 815 kcal/kg (3.3 MMBtu/ton) for flint glass with 60% cullet. Further improvements are expected with full integration of heat recovery systems. The performance of a net generation TCF design oxy-fuel fired container glass furnace with full heat recovery was estimated, based on actual experiences and model predictions.

REFERENCES

1. Kobayashi, H., "Advances in Oxy-Fuel Fired Glass Melting Technology", XX International Congress on Glass, Kyoto, Japan, September 26-October 1, 2004
2. Walsh, P.M., "Vaporization, Transport and Deposition of Sodium Vapor Species in Oxygen natural Gas Fired Soda-lime-silicate Glass Melting Furnaces", Twenty Eighth Symposium on Combustion, Edinburgh, Scotland, July 30-August 4, 2000.
3. Beerkens, Dr. R. G. C. and Kobayashi, H., "Volatilisation and Particulate Formation in Glass Furnaces," Advances in Fusion & Processing of Glass - 4th International Conference, - Wurzburg, Germany, 1995.
4. Faber, A.J. and R.G.C. Beerkens, "Corrosion of Combustion Chamber Refractories by Glass Furnace Atmosphere" TNO report . July 1995. Proprietary report to Praxair
5. Beerkens, R.G.C., M. van Kersbergen, O.S. Verheijen, "Experimental and Theoretical Simulation of Refractory Attach by Glassmelt Vapors in Natural Gas/Air and Natural Gas/Oxygen Fired Furnaces" TNO report . February 1997. Proprietary report to Praxair
6. Beerkens, R.G.C., and H. Kobayashi, "Refractory Exposure Tests and Modeling of Refractory Attach in Oxygen Fired Glass Furnace" July, 1998. Proprietary report to Praxair

7. Kobayashi, H., Wu, K.T. and Richter, W. "Numerical Modeling of Alkali Volatilization in Glass Furnaces and Applications For Oxy-Fuel Fired Furnace Design", Glastech. Ber. Glass Sci. Technol. 68 C2 (1995) pp119-127
8. Wu, K. T. and H. Kobayashi, "Three Dimensional Modeling of Alkali Volatilization / Crown Corrosion in Oxy-Fuel Fired Glass Furnaces", The 98th Annual Meeting of the American Ceramic Society, Indianapolis, IN, April14-17,
9. Böllert, J.E., "Heye oxy-fuel Technology – Principles, Layout, Production Experience", V Latin American Technical Symposium on Glass, Campos do Jordão-SP Brazil, Sep 21-25, 2003
10. U.S. Patent 5,924,858 (July 20,1999), "Staged Combustion Method" G.B. Tuson, R.W. Schroeder, and H. Kobayashi
11. U.S. Patent 5,628,809 (May 13, 1997) "Glass Melting Method With Reduced Volatilization of Alkali Species", H. Kobayashi
12. Allendorf, M.D., et. al., "Analytical Models for High Temperature Corrosion of Silica Refractories in Glass-Melting Furnaces", 7th International Conference in Advances in Fusion and Processing of Glass, Rochester, NY, July 27-31, 2003
13. U.S. Patent 6,253,578 (Jul. 3,2001), "Glass Melting Process and Apparatus with Reduced Emissions and Refractory Corrosion", H. Kobayashi and K.T. Wu
14. Snyder, W.J., Schroeder, R.W., and Wu, K.T., "Combining Oxy-Fuel Improvements for Maximum Advantage," Glassman 2000, Pittsburgh, PA, May 1 to 3, 2000.
15. Wu, K.T. and Kobayashi, H, "Comparative Evaluation of Luminous and Non-Luminous Oxy-Fuel Flames on Glass Furnace Heat Transfer," AFRC Spring Technical Meeting, Orlando, FL, May 6 to 7, 1996.
16. U.S. Patent 6,394,790 (May 28,2002), "Method of Deeply Staged Combustion", H. Kobayashi

A Designer's Insight Into All-Electric Melting

Carl W. Hibscher, (TECO), **Peter R. H. Davies,** (KTGSI),
Michael P. Davies, (TECO GLAS Ltd.) **and Douglas H. Davis,** (TECO)

Introduction

Fire and glass are inextricably bound, per the motto of the National Fraternity of Ceramic Engineers, "Through Fire to Perfection." Yet, an obscure glass property, electrical conductivity, has developed into an important commercial technology, i.e., melting glass with electricity. Glass that melts itself, without fire. Magic indeed!

Voelker's 1902 patents claimed feasibility, but little electric melting was done until the 1950's when pioneers showed that molybdenum electrodes lasted for extended periods in many glasses, without discoloring the glass. Initial developments were in the borosilicate and lead glass industries, both industries with serious emission problems. Glass fiber became a major producer in these all-electric melters.

For this article, we narrowly define "Electric Melter" as one that uses only electricity to heat glass, via Joule-heating. We will not cover electric boosting of fossil-fuel melters. This precise delivery of localized energy has seen widespread adoption, and the TECO group has supplied many boost systems. There are other all-electric systems such as induction, DC arc, plasma, or even microwave melting, which do not currently significantly influence the glass industry.

Types of All-Electric Melters

The majority of the successful all-electric melters can be broken into two categories. The most common, which can be designated as "Uniform Melters" have electrodes positioned and sized to provide reasonably uniform, symmetrical energy release, minimizing hot spots and convective flows from uneven heating. Most, but not all of these, are cold-tops as shown here, with a complete layer of raw batch insulating the surface (fig. 1). A flux-line diagram of a single square melter of this type with three vertical electrodes in each corner would indicate a reasonably uniform heat release in the glass, especially with vertical electrodes. Of course, the energy release will certainly be more concentrated at the elec-

Figure 1.

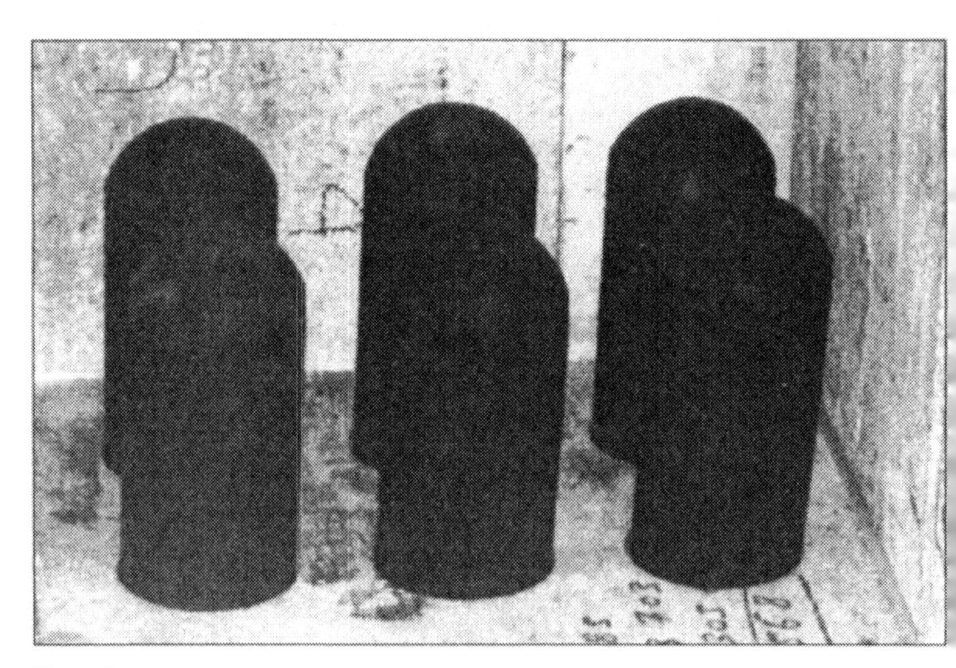

Figure 2.

trodes, and in fact math models of these furnaces confirm that there is still considerable flow in these melter. Therefore the term "uniform melter" best refers to the approach, not the end result.

These melters use molybdenum (Mo) or tin oxide (SnO_2) electrodes, depending mainly on the oxidation state of the glass. Tin oxide is suitable for oxidizing conditions, but its relatively low melting point restricts its use. It is either directly cooled or carefully placed to avoid peak glass temperatures. SnO_2 is used as massive stubs in the wall or bottom as shown here (fig. 2), or as wall sections of stacked blocks so that current density will be low and the SnO_2 can be directly cooled from the back. Tin oxide is used by the author's group almost exclusively in lead glass melters, where raw materials require oxidation.

Molybdenum, on the other hand, is rigid at high temperatures, but must be protected from oxidation. The high temperature strength of molybdenum allows it to be used as multiples of relatively small diameter rods, with long immersions to give high surface area; minimizing temperatures immediately around the electrode. Vertical, bottom-entry, or horizontal side-entry Mo electrodes are both used in cold-top melters; allowing flexibility in locating electrodes so as to give symmetrical, reasonably uniform energy release in the glass (fig. 3). The high surface area also allows higher energy input without high current densities that reduce electrode life and give hot spots. Improved electrode holders, using water-cooling for glass sealing and, depending on the glass seal expected, inert gas flushing in

Figure 3. Molybdenum electrudes.

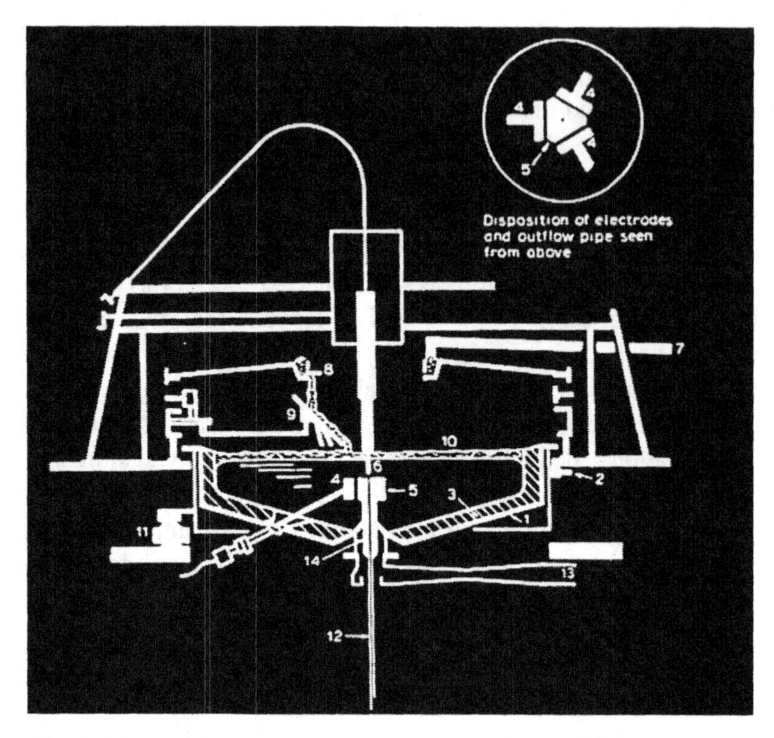

Figure 4.

the holder, make the use of bottom rods practical. These new style holders permit periodic, safe, electrode advancement, or even replacement on the run. Electrode diameters up to 4″ maximize campaign life. These uniform, cold top melters traditionally have only required collection of dust from the charging operation.

Platinum is used only for very special applications due to cost.

Another type of all-electric melter is the Intensive Melter (fig.4), where energy release is purposely concentrated in a small area, generating high temperatures and strong convective flows. These are related to the Pouchet-type melter, although the degree of energy concentration varies. These melters have been used for textile glass and in some insulation fiberglass facilities where their small size, low initial capital, and short rebuild times make their relatively high energy cost per ton of glass and short campaign life acceptable. High temperatures and intense flows assist in homogenization and dissolution, although they have short residence times. In most cases, they will require emission control, partly from charging dust, but also for volatilization of boron oxide et.al. from the extremely hot glass at the exit, and from the surface depending on the condition of the batch crust.

Electric Melter Ups and Downs

Figure 5.

As discussed above, in addition to lead glass applications, the borosilicate glasses and specifically glass fiber became a major market for electric melters, due to a need to reduce boron emissions. Further impetus was added in the 1970's, when natural gas and oil prices were rising, with no end in sight, due to reported dwindling gas sources. During this time, electric power was more stable, especially in the U.S. with abundant coal and availability of nuclear power. The efficiency of direct Joule-heating melting, lower capital for similar tonnage, and with low emissions, made electric melting a priority for many of our clients. In the twenty years from 1960 to 1980, TECO and its associated companies, KTGSI and TECO-GLAS Ltd., built over 100 all-electric melters.

In more recent years, energy preference has shifted away from electricity, largely due to new availability of low cost gas and oil and volatility of the electric markets. More recently, with the decreasing cost of oxygen generation, oxy-fuel melting has become an attractive alternative to electric melting. As a result, from 1980 to present, we have only built about 20 new electric melters. However, they are on the list we help clients choose from, and we expect to build more all-electrics - if the conditions are right for our clients.

Benefits of Electric Melters

In spite of relatively high electric costs in most areas, an electric melter uses that energy very efficiently. Heat is generated within the glass and is well-insulated by the batch blanket over the melt. Soda-lime and sodium borate glasses can be produced in an all-electric melter for 750-800 kWh/ton (2.5-2.7 MMBtu/ton), about half that commonly used on air/gas melters. This equates to roughly 80% fuel efficiency.

The cold-top batch cover, of course, also minimizes emissions. The sole escape route for batch decomposition gases and volatile species is through this blanket. Species (such as Na_2SO_4) will be condensed within the batch cover and carried down again to the melting interface. As a result, refining agents in a cold-top electric melter only need to be a fraction of those required in a fossil-fuel melter. Cold-top electric melters have been instrumental in reducing emissions from lead glass, and in hazardous waste/radioactive glass melting where the batch cover retains many dangerous elements volatilized from the glass. In the past, all-electric melting has been a complete answer to environmental compliance, except for simple dust control from batch charging. It recently has been necessary to provide more control of particulate and pollutants due to tightening regulations.

If "carbon taxes" become economically significant in efforts against global warming, electric melting could be more cost-effective, making use of untaxed "green" sources of power (hydro-electric, wind, biomass, solar, etc.) may increasingly be a practical and cost-effective glass-melting solution.

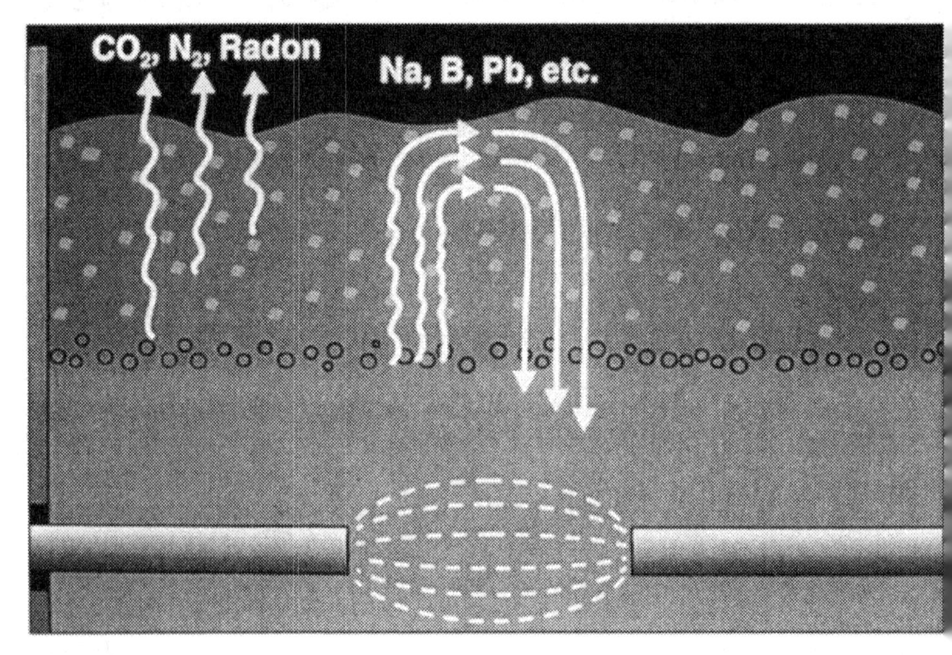

Figure 6.

Negatives of Electric Melters

Cold-top electric melters are not as simple to operate as fossil fuel. Finding a balance between blanket thickness, batch chemistry, glass temperature, and glass quality requires some counter-intuitive decisions, plus significant patience. Cold-top melters come with another variable, electrical resistivity, which varies with both temperature and batch chemistry. In electric melters, any change in pull, temperature, etc., must be done in slower, smaller steps than with a fossil-fuel melter. There is a tendency among first-time users of electric melters to excessively "tinker" with the process, making it difficult to ever reach equilibrium.

Cold-top electric melters do not allow for large variations in pull rate. As pull is reduced, glass temperature must also be reduced to maintain the batch blanket, and with lower temperature at some point glass quality (refining) will be a problem. Turndown to 50% of nominal would be great; turndown to 75% more likely. Fossil fuel melters have more flexibility.

We mentioned above that "most" of the "uniform" melters are cold-tops. Amber glass is difficult to manage with a cold-top, with serious foam generation. In the retrieved sample shown here, between the underlying glass and the raw batch, is a heavy foam that often results from hot spots within the dark glass and reboil sensitivity with some amber batches. It is often necessary to go to partial batch cover with some surface firing.

Very high levels of cullet in the batch can also be a problem in maintaining the batch blanket on a cold-top. It reduces the insulating power of the blanket, providing a window for escape of infra-red energy. High cullet levels from post-consumer sources also increase the odds of redox differences and associated foam.

Figure 7.

Glass plant profitability has generally been addressed by increasing melter size, spreading capital costs over more tons of glass. However, all-electric melters seemed to reach a practical maximum at about 300 tons/day. Many electric melters are commonly built on the "square pattern" with bottom entry Mo rods at the corners of the "square." Higher tonnages (up to about 300 tons/day) have been obtained with two "squares" put together with a single central throat. Even higher tonnage by combining more "squares" is logical, but there has not been a demand.

Campaign life of an electric melter is normally less than with a fossil fuel melter, often dictated by throat life. Average refractory temperatures are hotter and convection flows from concentrated heat release close to the electrode accelerates wear. Melter campaign life has improved due to development of new refractories. While earlier soda-lime and sodium borosilicates melters saw roughly 2 years life, later campaigns using chrome and fused zirconia (AZS) refractories have doubled to about 4 years.

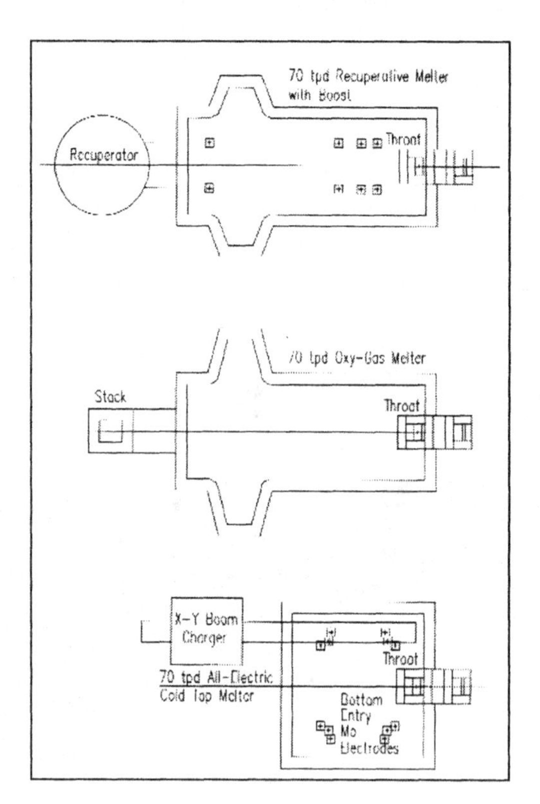

Figure 8.

Comparable Economics

A client came to us concerning a new 70 metric ton per day melter. The options considered were: (1) A gas-air melter with two-stage recuperator and significant boost, (2) An oxy-gas unit with no heat recovery and no boost, or (3) An all-electric cold-top melter (EMF). Location is always key, usually defining the energy costs.

The following reviews the effect on melting costs for the four current energy pricing situations shown here. The costs of natural gas and electrical power are current and actual. For the cost of oxygen, however, we chose to calculate the rate. The actual cost of oxygen will depend, of course, on the electricity cost, but also on supply volume, generation mode, local market situation, and vendor competition. For our purposes, we utilized the local power cost and added a "reasonable" markup to cover the return on VPSA units to the oxygen vendor for these small applications.

The accumulated cost of energy per ton of glass (fig. 9) for the three alternative melter types and four locations is shown graphically here. With California pricing for this small melter (left-hand three bars), the cost of natural gas and oxygen (no boost) for the oxy-gas melter should come to roughly \$39/ton of glass. The recuperative melter with electric boost should cost roughly \$41/ton for natural gas and electricity. The electric melt furnace (EMF) at the California site requires no natural gas or oxygen, but the cost of electricity alone totals

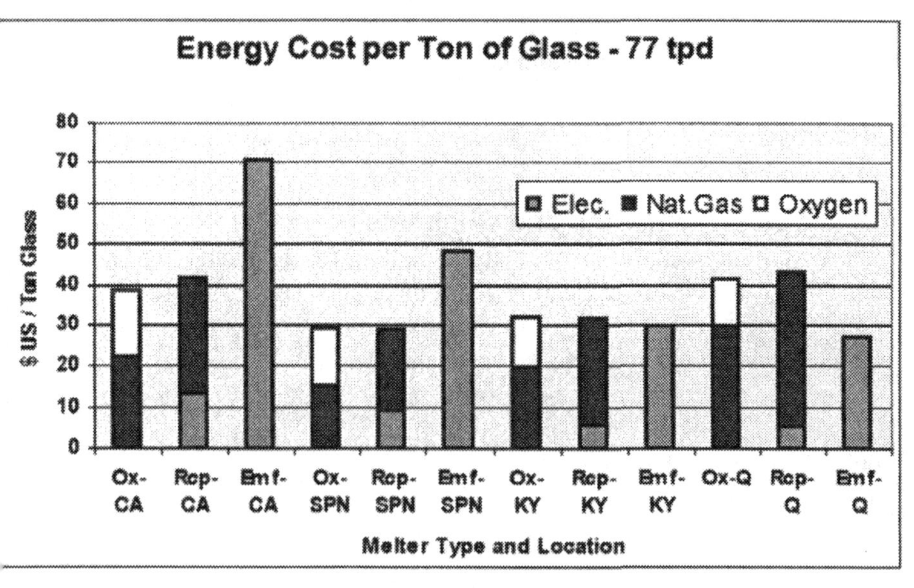

Figure 9.

$70/ton of glass. Obviously the EMF is a high cost choice here. In Spain, the oxy-fuel and the recuperative melter are essentially the same, while the EMF is again high cost. In Kentucky, the three choices are essentially the same. In Quebec, the EMF is clearly the lowest price choice.

A more detailed comparison was developed including capital and operating costs (fig.10). This covered a twelve-year period so that differing campaign lives and rebuild costs were taken into account. However, the added factors were essentially neutral and implications drawn from energy alone were the same.

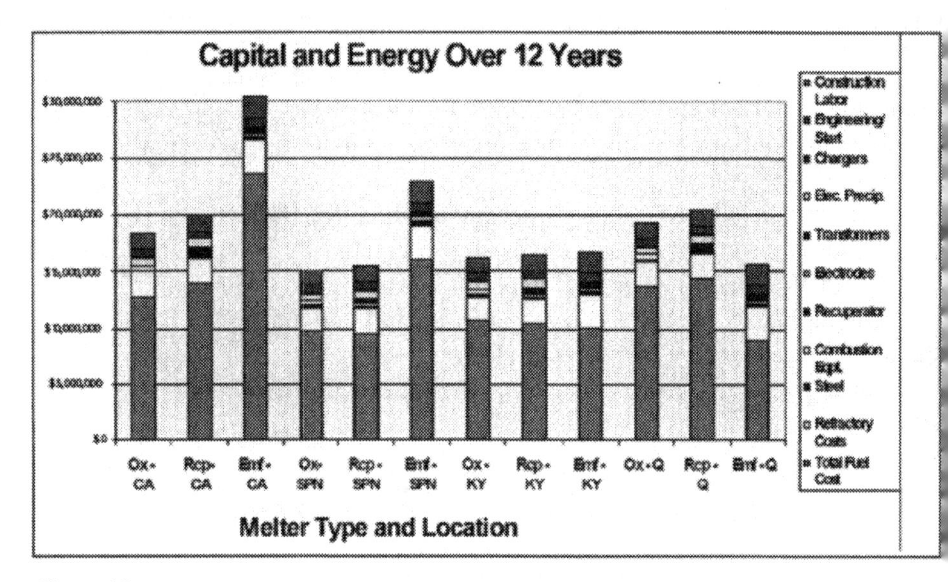

Figure 10.

All-Electric Forehearths and Channels

If the local price scenario favors electric melting for the melter, we should consider also using electrical heating for temperature control on the way to the forming operation. The energy cost normally resulting from the energy required in forehearths and channels can be surprising.

In an E-Glass (continuous fiber) operation (fig. 11), there is a substantial length of channels leading the molten, refined glass to the many individual bushings required to turn the glass into tons of fine fibers. As an example, one E-Glass operation uses 28 MM Btu/hr in their melter, and then in addition, another 17 MMBtu/hr to control the glass cooling and maintain that temperature for forming. Roughly 40% of the total energy bill is associated with the forehearths and channels.

Of course, the particular arrangement chosen was based on a complex mix of choices, including the size of fibers needed, number of bushings, the total tons of fiber needed, the arrangement of the fiber-forming operations below the melter, the number of furnaces, and more.

Another example is a wool fiberglass plant (C-glass) that has already made the choice of an all-electric melter (fig. 12). This melter operates at 5,850 kWh/hr, or 20 MM Btu/hr. However, for various reasons, the forehearths transporting the glass to the spinners are fired with natural gas, consuming 15 MMBtu/hr. Putting these on the same energy basis, the forehearth system constitutes 43 % of the total energy consumption.

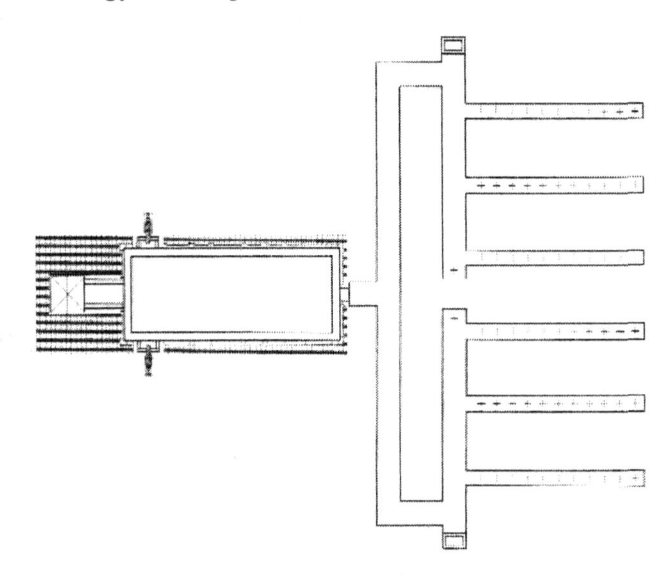

Figure 11.

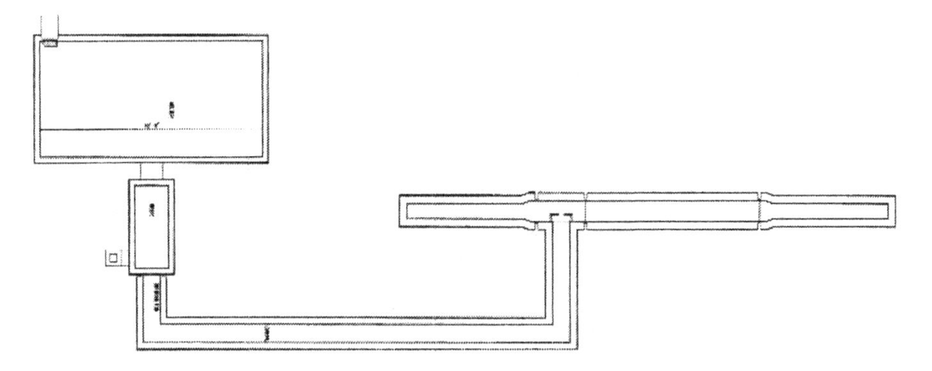

Figure 12.

Thus the energy required for controlling the glass past the melter is significant. If all-electric melting is one of the possibilities for your melter, or even if it is not, you should extend that consideration to the forehearths. The extremely high efficiency of joule-heating can produce overall saving even if electricity is more expensive on a straight energy comparison. All-electric forehearths also eliminate the large volume of combustion gases usually exhausted into the plant environment. While the low-temperature combustion does not generate significant amounts of NOx, both the heat and the gases are less than desirable.

Most of the glass industry applications can be done using all-electric heating. All-electric forehearths have taken several approaches. In most cases, a major part of the energy is applied from in-glass joule heating with electrodes of various materials being inserted into the glass and a potential applied. The energy is transferred directly to the glass by I2R heating of the glass itself.

We do hear concerns that exaggerated heat release at the tip of a worn electrode could result in severe local heating with reboil and attendant bubble defects. This is very uncommon, since the design takes this into account. First, the nominal amperage applied is normally limited to 10 amps/in2 or less for a normal reduced soda lime. For glasses with more reboil potential, the surface loading is kept even lower. Secondly, the operator monitors the amps and volt history of the electrodes, which will signal changes in the surface area being used for firing. In most cases, a worn electrode can be advanced further into the glass, or even replaced entirely on-the-fly. In a section of all-electric forehearth, the typical pattern is shown of single phase firing diagonally across a zone where electrodes directly across from each other are at the same potential. Thus advancing a worn electrode does not aggravate an already hot tip, because current passage is side-to-side, and not tip-to-tip between two electrodes (fig. 13).

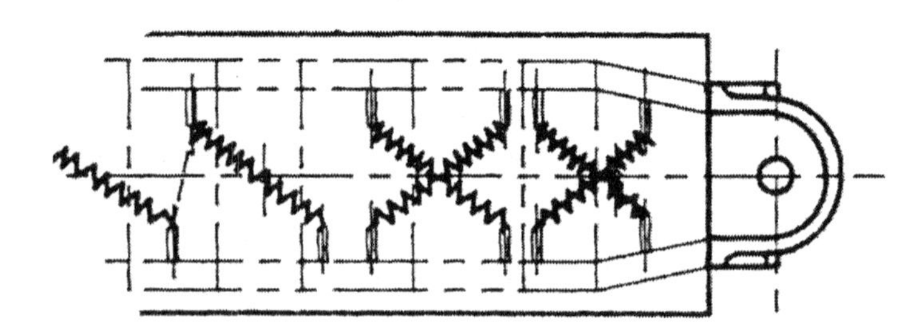

Figure 13.

Of course the goal of the forehearth, etc. is not only just to transport glass, but to bring this glass to a uniform desired forming temperature. This is a complex job. Additional electric heat has been used in the superstructure above the channels, via SiC and MoSi2 radiant elements. These can be used directly between the glass and the refractories, or with highly volatile glasses, with refractory muffle plates covering the glass, still allowing diffused radiant heating. In other cases, electrical elements have been installed within the refractories to slow the loss of heat from the glass, and provide an adjustable control to the effective "insulating power".

Probably the last forehearth situation to be tackled has been E-glass or other continuous fiber compositions, where the glass has extremely high electrical resistance due to low alkali. We have provided proposals on such systems and are convinced that the all-electric concept would work well.

Conclusion

The all-electric melter and forehearth is a viable commercial glass system in TECO's catalog. Currently, due to the cost of electricity and new alternatives such as oxy/fuel firing, the use of all-electric melters has declined. However, given the right pricing situation, they can be the correct tool.

Lessons Learned in Developing the Glass Furnace Model

Brian Golchert, Energy Systems Division, Argonne National Laboratory

Introduction

Over five years ago, the Department of Energy's Office of Industrial Technology funded a program to develop a user-friendly coupled glass furnace simulation (the Glass Furnace Model—GFM). The people involved with the code development (Argonne National Laboratory) were not experienced glass furnace modelers. This allowed for a different perspective on modeling glass furnaces. During the course of the project many furnaces were modeled, several new models were developed, and many interactions took place between the modelers and members of the glass industry. From these interactions, several practical engineering 'conclusions' were discovered concerning glass furnace physics and operations. This paper will discuss these practical aspects from an engineering point of view with very little emphasis on computational techniques or numerical models.

Simply put, a glass furnace takes energy to melt the batch/cullet, to heat the molten glass, and to refine the glass. Any energy that is not used for the above processes is a loss and thus lowers the overall furnace efficiency. In order to analyze the efficiency of a furnace, one can create computational simulations of the furnace that account for the transport of mass, momentum, energy, and species throughout the furnace. The field of computational fluid dynamics (CFD) encompasses the ideas of modeling physical process based on conservation of these key variables. CFD has been applied to modeling glass furnaces for almost two decades now.

Computationally, a furnace can be split into two regimes: a combustion space and a glass melt. The reason a furnace is split in two regions is that there are different types of flow present in each region: a fast flow (tens of meters per second) in the combustion space and a very slow flow (millimeters per second) in the melt space. These two different regimes make it difficult (but not impossible) to create one single simulation of the total furnace. An overall simulation is thus a combination of a combustion space simulation and a melt simulation. At the interface between these two regions, the heat flux calculated from the combustion space is used as a heat source for the melt code while the glass surface temperature calculated from the melt code is used as a boundary condition for the combustion space calculation.

So, from the start, modeling of a furnace makes an approximation to reality and it is these approximation(s) that force any modeler to provide the following caveat: computational fluid dynamics is not an exact science nor is it a 'predictive' tool. This statement is particularly true in the glass industry but requires further explanation. CFD is only as good as the quality of the input data and is only as good as the underlying physical models used to construct the simulation. Many glass furnaces either operate in steady state (no changes in the firing or pull rates) or in pseudo-steady state (regenerative furnaces). However, if one physically observes a furnace, it becomes apparent that 'things' are constantly changing (flames moving around, batch shape changing, etc.) even if the operation of the furnace does not change. When one models such a process, the modeler must consider these fluctuations when interpreting the computed results. In addition, the flow rates, material properties, thermophysical properties, pull rates, and even the geometric dimensions are not known precisely thus adding to the uncertainty of the results. What this implies is that a single computational model built for one specific geometric condition and one specific set of operating conditions is only directly applicable for those conditions. One can adjust the models to fit (as precisely as possible) whatever in furnace measurements are available for that set of conditions. However, this does not imply that the modeler could then use this finely tuned model to simulation another set of conditions and expect to exactly match any measurements from those new conditions. This is what is meant by the statement "CFD is not a predictive tool."

The utility of CFD becomes apparent when analyzing two sets of similar conditions. For illustrative purposed, assume the first case is the 'normal' operating conditions and the second case represents a change in the firing pattern for the same furnace. Comparison of two simulations will allow the user to see if any trends have formed (energy transferred to the melt increased, NO_x decreased, wall temperatures decreased, etc.) based on the change in firing pattern. While it may not be able to precisely predict furnace conditions with the new firing pattern, CFD is an excellent tool to determine the trends that were caused by this change in firing pattern. The word 'precisely' was put in quotes since most CFD codes are able to 'predict' reasonably well (proper order of magnitude if not relatively close to actual measurements) the conditions in the furnace for the new firing pattern. However, it must be emphasized again that CFD simulations will allow the user to determine the trends between similar simulations and thus allow the user to use engineering judgment to determine how to optimize furnace performance.

In regard to the quality of the models used, many of the basic phenomena concerning combustion and glass flow are relatively well known and the choice of the models depends on the level of detail that is of interest to the user (for example, seven chemical reactions as opposed to three hundred reactions). However several key processes vital to glass furnace modeling are not fully known or understood (for example, chemical reactions in the melt and soot formation/oxidation). The lack of uncertainty in these models is of lesser importance since one is comparing two cases that have the same set of physical models present. Any errors caused by these models would thus be present in both simulations and would therefore not affect the trends present.

Energy Transport in the Combustion Space

In the combustion space, fuel and oxidizer mix, combust and then transfer the released energy. Ideally, one would like as much of this energy as possible going into the glass melt but a portion of the energy goes through the crown and a lot goes out the flues. Proper placement of burners and exhausts as well as well-designed firing patterns can improve the efficiency of a furnace.

From the extensive number of furnaces modeled throughout the project, it has become apparent that the location of the flues with respect to the location of the burners is not optimal with many of the non-regenerative types of furnaces, in particular with the cross fired oxy-fueled furnaces. In many cases, one or more of the flues will be running at a much higher temperature than the other flues. This is caused by one of the burners 'short circuiting.' The heat released from this burner does not have sufficient time to be transferred to the melt before it leaves the furnace. In fact, if one closes the exhaust that is running at a higher temperature, the heat transfer to the glass will increase. Of course, closing an exhaust may not be physically possible in an existing furnace but it does give one information to consider when designing a new furnace. Figure 1a and 1b shows the temperature field in the plane of the burners for a cross-fired oxy-fueled furnace with two flues located in the batch region. Figure 1a shows the normal operating conditions while Figure 1b closes the exhaust with the higher temperature. As can be seen in the figures, the temperatures throughout the furnace increase by closing the one exhaust. The ensuing increase in gas temperature increased the total heat transferred to the load without increasing the firing rate.

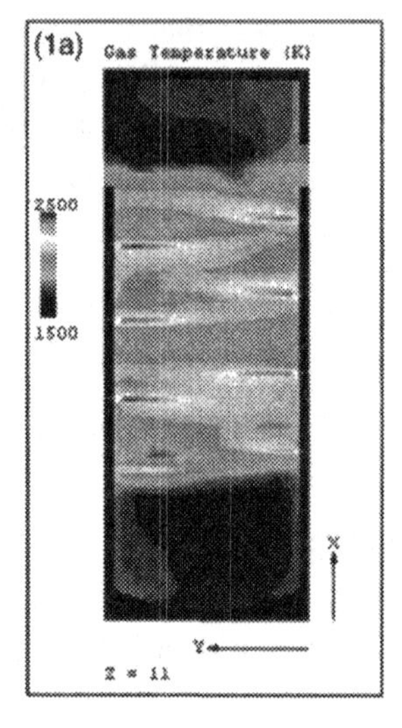

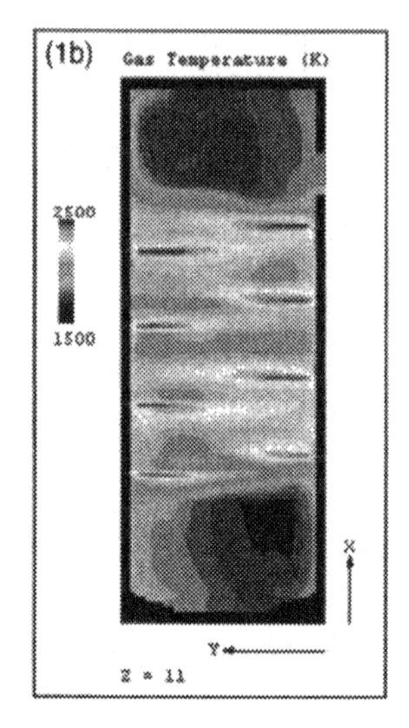

Figure 1a and 1b. Gas temperatures under normal operating conditions (1a) and with one exhaust closed (1b)

In order to optimize performance, a modeler needs to ask the question: "Where does the energy come from?" This does not mean what type of fuel but what physical mechanism dominates or controls the heat transfer from the combustion space to the melt. There are three main modes of heat transfer: conduction, convection, and radiation. For these high temperature environments, radiation dominates. Thermal radiative heat transfer can also be broken down into two main components: gaseous radiation (mainly carbon dioxide and water vapor) and soot radiation. Of these two processes, soot radiation is the controlling process, particularly for oxy-fueled furnaces. Soot is created in fuel rich regions where the local temperature reaches a high enough temperature to permit a small amount of cracking (breaking off of parts of the larger hydrocarbons). Once the soot is created, it is transported by the gas flow field in the combustion space. If the soot and local oxygen concentrations are larger enough in the same region, then the soot will oxidize (burn). When one looks into a furnace and 'sees' a flame, one is actually looking at the local soot oxidation.

The GFM was used to model a typical glass furnace and the energy released from soot radiation and from gas radiation was quantified as a function of wavelength (see Figure 2). As can be seen for each waveband, the amount of energy from soot is greater than the energy from gases except in the narrow band near the natural radiation radiating wavelengths. What this implies is that if one can control the formation and destruction of soot, then one can have a greater degree of control of the heat transfer from the combustion space to the melt.

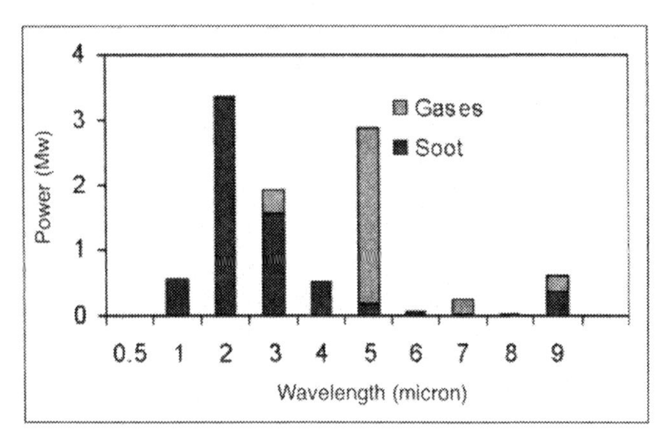

Figure 2. Energy released as a function of wavelength and source.

In part, this explains the efficiency of tube-in-tube oxy-fueled burners (and all of their variants). After many years of trial and error, a typical oxy-fuel burner involved a tube-in-tube arrangement. Fuel (natural gas) would be injected in an inner tube while the relatively pure oxygen would be injected in the outer tube (see Figure 3). This design helped ensure the complete combustion of the fuel, thus supposedly saving on energy costs.

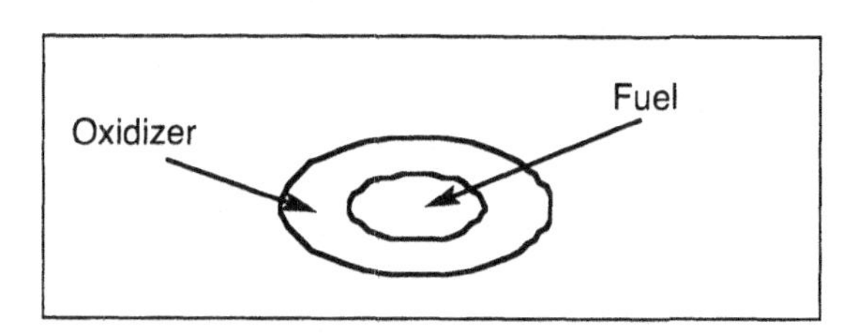

Figure 3. Basic shape for tube-in-tube burner.

The fuel near the oxidizer would combust, raising the local temperature but the fuel in the central portion of the fuel region would not combust immediately. As the temperatures elevated, more soot was formed due to the cracking of the un-burnt fuel thus increasing the radiative heat transfer. Eventually all of the fuel would be consumed. At this point, there would still be excess oxygen present in the outer regions. As the soot was transported through this oxygen rich region at these elevated temperatures, much of the soot would oxidize. Thus, these oxy-fuel burners would allow greater control of the local generation of soot and, in turn, allow better heat transfer to the melt.

In addition to the radiation from combustion, one must consider the re-radiation from the crown. Since the crown is a diffuse emitter (sends radiation in every direction), it has the effect of spreading the heat flux pattern generated by the flames. It is an important contribution to the total heat transfer in the furnace as can be seen in Figure 4. From this figure, one can see that the contribution to the total surface heat flux from the wall is on the same order of magnitude as the energy from the flames. Also on Figure 4 is a line indicating an approximate glass transmissivity. The larger the values on the green line, the greater the chances that the thermal radiation will penetrate into the liquid glass. This is an important consideration when considering the Rosseland approximation for radiative heat transfer in the melt.

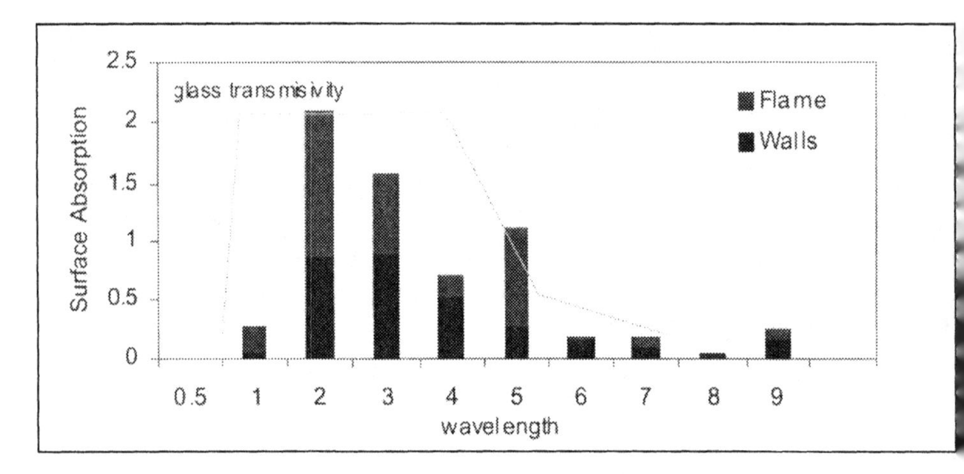

Figure 4. Contribution to the surface heat flux from the flames and from the walls

Effect of Surface Heat Flux on Melt Flow Field

Once one understands the physics behind the heat transfer from the combustion space to the melt, the next logical question would be: "What does the surface heat flux do?" How does the magnitude and the location of the surface heat flux affect the movement of the molten glass? Aside from electric heating of the melt, the surface heat flux provided the energy needed to heat the batch, to fuse the batch, and to heat the molten glass in addition to compensating for heat losses through the walls. After modeling a number of different types of melters, a few 'generic' recommendations concerning the distribution of the surface heat flux can be made. Again, these are general conclusions and that each furnace should be modeled and examined individually.

Depending on the distribution of the surface heat flux, one can have backflow through the throat of the melter (assuming a refiner is present). Instead of having all of the liquid glass moving in the same, outward direction in the throat, some of the glass actually flows in the opposite direction (back towards the chargers). This backflow effect has been seen in several models of melters that have been created. Backflow is caused by the liquid glass in the refiner being cooler than the liquid glass in the melter (near the throat). Since the refiner glass is cooler, it becomes more dense. As it's density increases, it 'sinks' to the bottom of the refiner and sometimes pushes its way back through the throat. Whether it can actually push back into the melter depends on the surface heat flux distribution in the melter.

If the surface heat flux is higher towards the throat, the glass near the throat naturally heats up. As the glass heats up, the viscosity decreases and the glass flows more readily. This hotter surface glass flows in two main directions: towards the batch and towards the backwall. As the hot glass hits the backwall, it is forced to go downwards towards the throat. In many cases, this hotter glass then moves directly through the throat into the refiner. If this hotter glass encounters cooler glass coming down the refiner, it usually has enough energy to turn the backflow around.

On the other hand, if the surface heat flux is more uniformly distributed over the glass surface, the surface glass near the throat does not heat up as much and moves less quickly. Since this short-circuiting glass has less energy to stop the movement of the cooler glass from the refiner, backflow may be established. Figure 5 shows two cases. 5a shows a case with more uniform heat flux while 5b has the heat flux peaked towards the back wall. Figure 5a shows backflow while 5b does not. Clearly, how you distribute your heat flux (which depends on how your fuel is burned) dramatically affects the flow field in the melt.

When backflow is present, the glass does not immediately leave the refiner into the forehearths. This lengthens the time the liquid glass resides in the melter/refiner and should improve the glass quality.

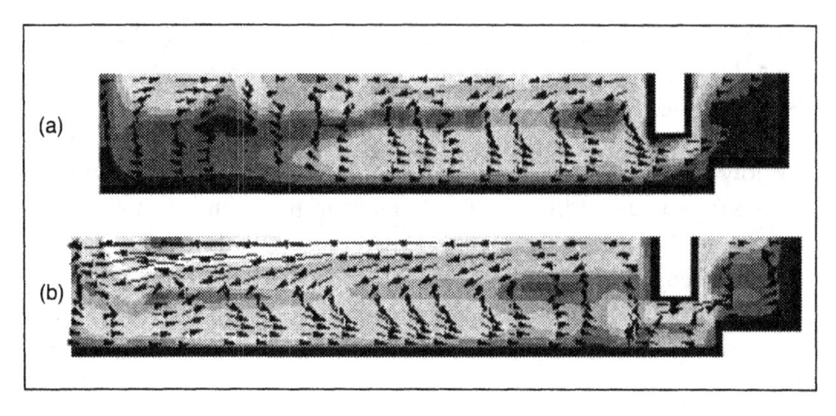

Figure 5. Melt flow fields for heat flux peaked near back wall (a) and more uniform heat flux (b).

What can affect the transfer of energy from the combustion space to the melt? A long held industry belief is that foam is a thermal insulator. Many glass melts have foam covering portions of the glass surface. Foam is created by chemical reactions in the batch and in the melt. Batch reactions release gas with the majority of those gases immediately being released into the combustion space. However, a portion of these gases are entrained in the melt as bubbles. Bubbles are also created by chemical reactions (fining) in the molten glass. The transport and growth of these bubbles can be modeled using the GFM. Once the bubbles reach the surface of the glass, they can burst or they can create a foam layer. Using the multiphase models in the glass melt portion of the GFM, a foam layer can be predicted (see Figure 6).

How does the foam affect heat transfer to the melt? From the physics, it can reflect, transmit, or absorb the radiation. Studies done at Purdue have shown that for typical foam thicknesses and typical bubbles sizes, approximately 40% of the incident radiation is reflected. This means that 60% is either absorbed or transmitted but in either case, 60% of the energy is still deposited locally. Clearly then, foam is not a perfect insulator. Of the 40% that is reflected, a small portion is absorbed in the combustion gases (raising the flue temperatures) but the majority of the reflected energy is absorbed in the crown. This raises the crown temperature and increases the energy re-radiated but is a diffuse fashion. Most of this re-radiated energy passes through the combustion gases and hits the glass surface where 60% of this energy is absorbed and 40% is reflected. This process keeps repeating and the net effect is that the surface heat flux is more spread out for the case with foam. In reality, foam becomes a radiation diffuser not a thermal insulator.

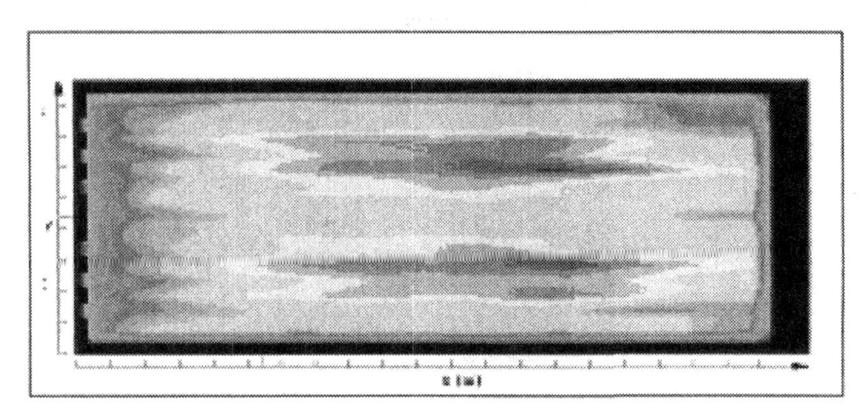

Figure 6. Computed foam distribution.

Conclusion

The purpose of this paper was to present some of the understanding of the transport of energy in a glass furnace. By better understanding how the energy moves about in the furnace, a furnace operator or designer can improve the performance (both in energy use and in quality of glass) of the furnace.

The SORG® VSM® All-Electric Melter for 180 t/d Container Glass—Design, Installation and First Experience

Matthias Lindig, Nikolaus Sorg GmbH, Germany, and
Jan De Wind, Consol Glass, South Africa

General

The Consol Group is the leading South African glass container producer with manufacturing facilities in Wadeville and Clayville near Johannesburg, in Pretoria and in Bellville near Cape Town. The company produces all kinds of containers, including a large variety of different glass colours, ranging from flint to light and standard emerald green, dark green, amber and blue. Color changes on the fly are made on most of the furnaces.

In 2002 SORG®, was asked by the Consol Glass Group to design a large all-electric melter for container glass.

The availability of energy sources in South Africa differs significantly to those in most other countries. Much use is made of their large deposits of coal since there are no indigenous oil reserves.

Coal is the main energy source for electric power generation, and it is also used to produce fuel gas. The gasification process has been known since the early 19th century, and in the early 1920s the Germans were processing coal in order to obtain gas and fuel. However, the process was never economically successful in the western world. Nevertheless, It is interesting to note that China has recently purchased a test facility from the Deutsche Montan Technologie, Germany to produce gas from coal. The Chinese believe it will be possible to utilize their extensive coal deposits to produce large quantities of gas in the near future.

In South Africa, the energy supplier Sasol has been supplying communities with gas using the so-called advanced Fischer-Tropsch process since the early 1950s. The gas produced is a reasonable and competitive alternative to oil due to the low wage levels. The gas has a remarkable hydrogen content, and a calorific value of 39 MJ/m^3.

As a result of the specific conditions relating to energy supply in South Africa the operation for all-electric furnaces is more economic than in most other countries. This is the reason that Consol Glass decided to build an all-electric melter for their new furnace project in 2002.

New furnace project, design and installation

Consol Glass gave SORG® the order for the design, supply and installation of the new furnace to be built at their Wadeville facility. The furnace was to be capable of melting up to 180 t/d flint glass, but was also to be capable of melting emerald green glass. The cullet ratio was expected to be 30%. The new furnace was to supply two production lines.

The standard SORG®, design of all-electric melter, designated the VSM®, is a cold-top, vertical melter. The batch is spread over the complete surface of the furnace and the glass is extracted through the throat at the bottom of the furnace. Thus, the melting and refining take place in a vertical direction rather than on a generally horizontal one as is the case in conventional furnaces.

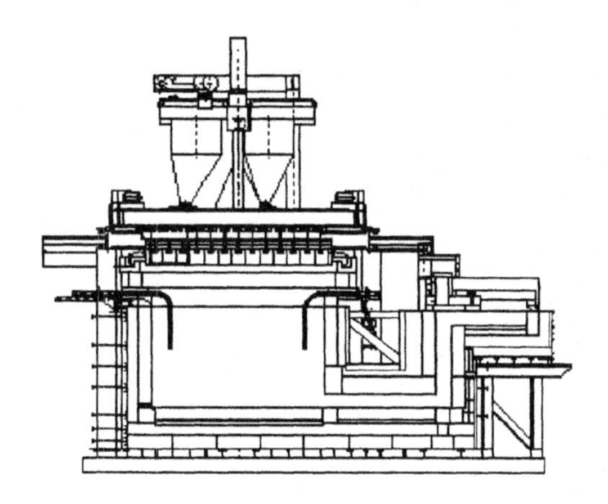

Figure 1. Longitudinal section of a SORG® VSM®, all-electric melter with top elec-trodes and rotating crown.

The batch charging is carried out using a rotating crown system. A number of vibratory feeders are mounted on top of the crown and these feed batch into the furnace through small openings in the crown which are arranged radially. During the batch feeding process the complete crown assembly, including the feeding units, rotates, and the feeders lay a series of concentric rings of batch onto the glass bath surface. The small feed openings in the crown can be easily sealed, and there is no need for any opening in the superstructure side wall for the batch charging.

This arrangement gives a completely enclosed superstructure, whilst giving excellent control of the batch blanket. If complete emission control is required, a small exhauster can be connected to the superstructure and the limited quantity of gases removed in this way can be treated in a simple bag filter.

A second important feature of the VSM® furnace design are the patented SORG® Top Electrodes. Electrodes are normally installed through the tank side walls—an arrangement which leads to increased side wall refractory wear. SORG® Top Electrodes enter the furnace through the superstructure side walls and pass into the glass bath vertically through the glass surface. A special holder design makes this possible. No holes and water-cooled holders are needed in the tank side walls—an important factor for reducing the side wall wear.

Although a range of glasses can be melted in this type of furnace, including normal soda lime glass, the relatively high cost of electrical energy in many countries has tended to limit the application of all-electric melting to a number of special glasses. Soft borosilicate C glass and fluoride opal glass are two of the glasses for which the VSM® electric furnace design is especially well suited.

More than 80 VSM® furnaces have been built, with design melting capacities from a few tonnes per day up to 150 t/24h, whereby the 150 ton furnace is used to melt C glass for insulating wool fiber.

The Consol furnace has a melting area of 80 m^2, whilst the tank depth is about 2,6 m. The furnace is equipped with 12 pairs of Top Electrodes supplied by two 3-phase transformers. The installed power is 9000 kVA. In the beginning the furnace started with 2 different lengths of electrodes, with 900 mm and 700 mm immersed into the glass. Eight charging slots were provided along the diagonal due to the large melting area.

The heating-up and beginning of the first campaign was in February 2003. The campaign started with flint glass. The pull was kept stable at about 182 t/d and the power input was about 6900 kW. The specific energy consumption was about 0,89 kWh/kg glass. The temperature in the superstructure was below 350 °C, glass temperature measured at the side wall was about 1500 °C and the glass temperature at the top of the riser was 1350 °C.

The glass quality was very good from the beginning. Shortly after the start-up the glass colour had to be changed from flint to emerald green, and the operator has had to face a number of problems since then.

Melting Problem Description and Countermeasures

From the time the glass colour was changed to green a change in the batch blanket was observed. The glass below the batch was boiling, whilst the batch layer itself appeared to be frozen, hindering the release of gas. From time to time the surface skin was lifted by the gas and broke up, releasing inflammable gas. It became difficult to control the batch layer and to keep the surface even and quiet.

A number of counter-measures were applied in order to stabilize the melting conditions.

The first measure was to increase the energy input. The surface skin disappeared and the batch layer became thinner and more controlled. However, this solution was not acceptable since the energy consumption increased, the entire furnace temperature was raised and the production was suffering from an unacceptable high glass temperature at the top of the riser. The forehearth cooling system was not able to cool down the incoming glass sufficiently.

Another measure was to add sodium chloride, which acts as a flux and helps early melting. However, the salt addition caused major air pollution problems and also resulted in rust attack on the molds and the machine. Fluorspar was added as an alternative as it works in a similar way, but it also suffers from the same disadvantages.

From the beginning it was thought that the problem was basically caused by the sulphur addition even though the same amount was used for the flint glass.

Changes were made to the batch charging in order to improve the surface conditions. The batch was charged more evenly in order to avoid open areas, so that the temperature below the batch would also be more even and surface skin would be avoided. Tests were also made in the opposite direction, with small areas being kept free of batch to help the gases release. However, none of the changes to the charging pattern resulted in an improvement.

The batch raw materials were rechecked. The sand grain size distribution did not include too many fines and there were no unusual impurities in the remaining batch components. In the end the major concern was the addition of sulphur. Obviously there was a difference in the gaseous sulphur oxide release between flint and green glasses, although the total amount of this gaseous species is small compared with the total amount of batch gases (about 1800 m^3/h CO_2 vs. 7 m^3/h SO^3).

In further tests involving modifications to the chemistry, the total amount of sulphur was reduced by half and ferrophos was added in an attempt to produce more reducing conditions and to avoid temporary readsorption of sulphur. Both measures helped to improve the conditions. However, the ferrophos had to be removed because of glass colour instabilities.

Problem Analysis and Explanation

Experience has shown that the behavior of sulphur in the green glass is different to that in the flint glass. The major difference in the composition of the glasses is the amount and number of polyvalent colouring oxides. Obviously the redox state of chromium, iron and sulphur is affected in the lower temperature regime below the batch blanket. With the addition of ferrophos the glass is more highly reduced, which results in a lower SO_2 solubility. It appeared possible that the SO_2 bubbles released in the hot area were ascending into the colder area and interacting with the glass.

A number of papers have been published that address the behaviour of sulphur in glass in general terms. In most of these contributions the redox state of the oxides is investigated under heating-up.[1-4] The main subjects of most papers are the redox equilibrium and the fining potential.

Looking at the situation from the other side, as the temperature reduces the redox potential between chromium, iron and sulphur results in reduction of the iron and dissolution of sulphur. The layer below the batch becomes increasingly saturated with sulphur, whilst the glass layer below the batch becomes reduced and less transparent to heat radiation. This phenomena might result in a surface skin impermeable to gaseous components. The more reducing the conditions, i.e. as the ratio of Fe^{2+}/Fe_{total} increases, so the solubility of sulphur decreases. This effect is well known and is described by Williams.[5]

In a paper by Müller-Simon[6] the thermodynamics of the polyvalent ions and the influence of the oxygen partial pressure on the sulphur solubility are discussed for flint, amber and green glasses. The dissolved oxygen is assumed to be independent of the temperature. The oxygen partial pressure and the concentration of the polyvalent elements can be used to calculate this value. The S^{4+} concentration is calculated relative to temperature for the different glasses.

The author came to the conclusion that the formation of S^{4+} is significantly different in flint glass compared to green glass, as shown in a comparison of figures 2 and 3. In the case of flint glass the formation of S^{4+} is almost stable at higher temperatures. However, in green glass there is a much steeper increase in S^{4+} formation due to the redox reaction between iron and chromium. The redox reactions are given in figure 4.

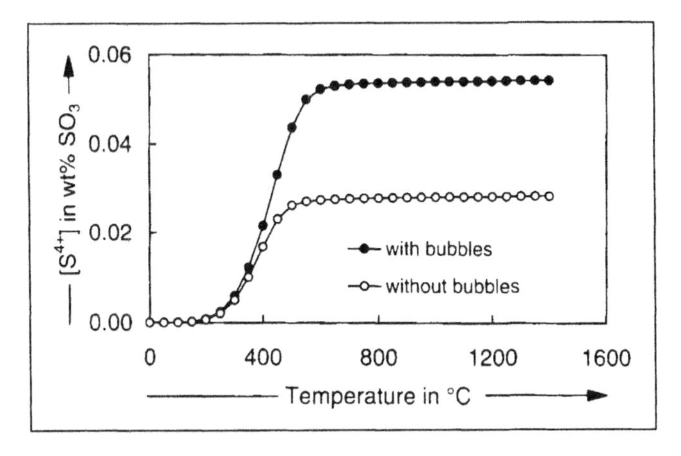

Figure 2. S^{4+} concentration vs. temperature for a flint glass melt.[6]

In the case of green glass there is a much higher probability of redissolution of the S^{4+}. The redox reactions make the glass more reduced, which change the conditions for the sulphur again. The solubility drops and degassing occurs near the glass surface. This is what the operator can observe from time to time.

Based on this result the quantity of salt cake used should be reduced as much as possible. It is usually sufficient to use one third of the quantity normally used in a conventional gas fired furnace.

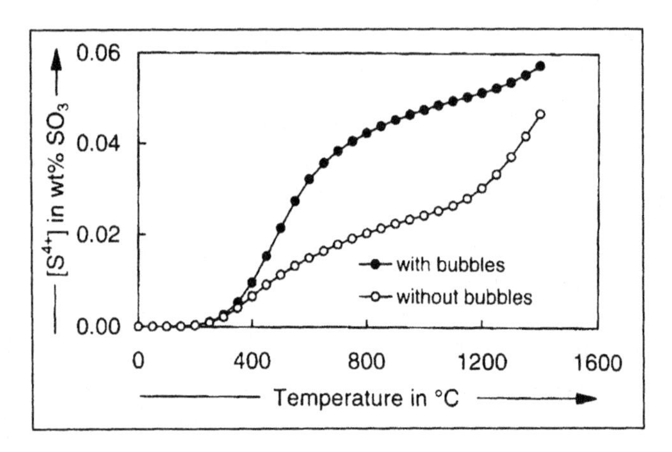

Figure 3. S^{4+} concentration vs. temperature for a green glass melt.[6]

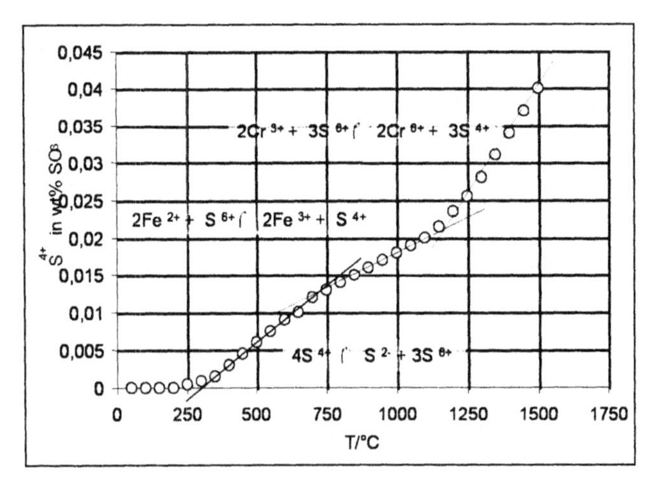

Figure 4. Variation of the sulphite concentration vs. temperature for a flint and green glass.

Experience has shown that the furnace load has an impact on the sulphur dissolution and release. At 160 t/d the furnace can be operated without problems. This is equivalent to a specific load of 2 t/m^2d.

In a paper presented to the German Society of Glass Technology (DGG) (7) a very similar phenomenon was described in an all-electric melter for lead crystal glass. The glass below the batch boiled from time to time. The problem was solved by adding niter and increasing the energy input. It was also assumed that the problem was caused by the $Fe^{2+/3+}$ conversion near the surface.

The addition of niter might ensure that oxidizing conditions exist in the layer below the batch, but since sodium nitrate decomposes at a low temperature, it would be preferable to use potassium nitrate.

In conclusion the vertical all-electric melter cannot be used to melt green glass at the same pull level as flint glass. The sensitive redox reactions require higher temperatures in the batch layer. The same result has also been found on other all-electric melters running with emerald green glass.

REFERENCES

1. Jebsen-Marwedel, H; Brückner, R.: *Glastechnische Fabrikationsfehler;* © Springer-Verlag Berlin Heidelberg New York, 1980, pages 210-214
2. Müller-Simon, H.: "On the interaction between oxygen, iron and sulphur in industrial glass melts"; Glastechnische Berichte (Glass Science & Technology), 67 (1994) 11, pages 297-303
3. Rudolph, S.: "Reduction of sulphur in sodium-silicate-glasses"; paper given at a meeting of the German Society of Glass Technology, October 21, 1986, Würz-burg/Germany
4. Müller-Simon, H.: "Temperature dependence of the redox state of iron and sulphur in amber glass melts"; Glastechnische Berichte (Glass Science & Technology), 70 (1997), pages 389-391
5. Williams, H.P.: "Influence of the redox state of the batch on the fining with sulphate"; Glastechnische Berichte (Glass Science & Technology), 53 (1980) 7, pages 189-194
6. Müller-Simon, H.: "Oxygen balance in sulphur containing glass melts"; Glastechni-sche Berichte (Glass Science & Technology), 71 (1998) 6, pages 157-165
7. Langer, A.: "Influence of the Redox State of the Glass Melt on the Operating Pattern of an All-Electric Melter"; paper given at a meeting of the German Society of Glass Technology, October 21, 1993, Würzburg/Germany

ACKNOWLEDGEMENT

SORG® gratefully acknowledges the co-operation of Consol Glass, South Africa, and in particular the open discussions that have taken place and the support provided. In addition we greatly appreciate the approval of Consol Glass for the publication of this paper.

The U.S. Glass Industry—Moving in New Directions

Michael Greenman, Glass Manufacturing Industry Council

Abstract

Since its formation in 1998, the direction the glass industry has been taking through the leadership of the members of the GMIC has been in line with the mission of the DOE, our primary partner: the reduction of energy use in our melting operations. As our collaboration has evolved, we are beginning to explore together a wide range of new initiatives: glass properties and qualities; raw materials, cheaper energy sources, communication with other regulatory government agencies, new forming methods, new funding mechanisms, innovative uses and types of glass. This paper will describe activities and status of these new initiatives.

Background

The Glass Manufacturing Industry Council (GMIC) was born out of the desire of its founding members to create an organization that would serve to represent the interests and priorities of the United States Glass Industry to the Department of Energy. The DOE was developing relationships with each of the energy-intensive industries it had identified as the "Industries of the Future" (IOF): Agriculture, Aluminum, Chemicals, Forest Products, Mining, Metalcasting, Petroleum, Steel and Glass. The IOF program has the objective of financially supporting cost-shared research with these nine industries with the objective of developing technologies that would reduce the energy intensiveness of these "Process Industries". Reducing energy use has always been of interest to the glass industry, so we gladly entered into a relationship with the DOE and created the GMIC to facilitate the communications and collaboration within the industry and with the DOE.

Fifteen projects have been completed under this program, covering all four focus areas that were identified by the glass industry as needing further study: Energy Efficiency, Production Efficiency, Environmental Issues, and Innovative Uses. The majority have been completed, their results quantified and hardware, software, or knowledge developed have been disseminated and implemented. Several are still in process, and rapidly moving towards completion. In another series of research projects, conducted within the "G Plus" program, almost 30 relatively small projects have been carried out in cooperation between GMIC members and the National Labs. These short-term projects have addressed a number of specific challenges that the companies have faced, or have answered questions that required the high level of expertise and sophisticated equipment that exists only at some of our labs. Solutions have contributed to our understanding of glass and its melting and forming.

All research conducted with the support of the Department of Energy, however, has been required, if it was to be supported, to lead to the likelihood of an energy savings in the melting process. Reducing energy intensiveness in manufacturing, after all, is their mission.

But is the reduction of energy use in our furnaces the only need we have if we are to improve the long-term viability of our industry? Clearly, there are many other areas of need: basic materials knowledge, lower capital costs, cheaper energy sources, new uses and applications for glass, and many others.

The addition to our staff in 2002 of a Technical Director, John Brown, formerly of Corning, has given us a tremendous boost in our ability to undertake new initiatives. I'd like to outline a few of the areas we are focused on, any one of which could bring substantial change and benefits to our industry if carried to successful conclusion:

Strength of Glass

Coming into the glass industry for the first time in 1995 I began to hear about the "Holy Grail" of glass: the search for a means to formulate, melt and form glass such that it would exhibit a tensile strength somewhere closer to its theoretical values of 1,000,000 pounds per square inch, than to the current limits of 7,000 psi, at least for glass containers. If the useful strength of glass could be increased to even 10-20% of the theoretical strength of glass, this would have a profound impact on the entire glass industry. The weight/strength performance and cost/strength performance of glass would be far above most common competing materials. This would greatly expand the markets in which glass could be used.

Attempts to increase the useable strength of glass are nothing new. However, it is felt that a fundamental study involving both theoretical and experimental components and aimed at understanding the formation of glass surfaces at typical glass processing temperatures offers the best hope of determining if and how it might be possible to either; 1) alter the surface of glasses so that their inherent strength is similar to the strength in the interior of the glass or; 2) to protect the freshly formed surface from attack by various molecules such as water.

So, We've initiated a two-stage process to pursue this dream: First, currently under way, is a contest to solicit ideas of how "strong glass" could be utilized

This is a collaboration between the GMIC and the "International Materials Institute on New Functionality in Glass". It will be announced to undergraduate and graduate students around the country in the near future. It will go, not only to students in glass or ceramics areas, but also in areas such as mechanical engineering, civil engineering, and architecture. It will request 1-3 page proposals consisting of specific applications and markets that could be envisaged if such "strong glass" could be made. Cash prizes will be offered, and a jury of respected glass scientists will pick the winning papers.

The second stage, will use the outcome of the first to sell the idea of what benefit a strong glass material would be to the economy as a whole and the revolutionary new products that could result from it. Planning for this stage is in process at this time and we are seeking funding from various sources to support a major contest, open to all interested individuals or groups, and involving academia, national labs and contract research organizations.

From this, with luck, focus, and determination, we believe will come the next generations of glass products.

Raw Materials

Over the years glass scientists have found new applications for glass by experimentation with the use of new materials in the batch: Early in this Century, "Tin Oxides" were added to insulator glass to reduce radio interference; Adding Phosphorus led to opalized or Phosphorus Glass and to bone replacement materials; The use of Neodymium led to the creation of high energy laser glass. Many such evolutionary steps came from interactions between glass artists and glass scientists. We will be exploring this process in Corning and Alfred New York in November when we bring together 7 glass artists and 7 glass technologists to discuss the possibility of collaboration aimed at moving both sides of the continuum forward to greater success.

Related to this effort is a new initiative that saw its onset on Monday afternoon in the form of our "Growing the Glass Market" workshop. This was the creation of an alliance between the raw materials suppliers, I represented by the Industrial Minerals Association, and the glass industry to begin to look at how our two industries can work together to advance the use and value of glass in the United States. Are the raw materials we use in our industry the optimum products for our needs? Are there other materials that could be proposed that might improve our glass quality, characteristics or cost-effectiveness? Lithium has been a topic of discussion as a possibly beneficial additive for many years. Could new melting technologies lead to opportunities to use new raw materials, or more flexible specifications?

The discussion was lively, and a wide variety of new ideas were developed which we'll be reviewing and discussing to seek a way forward to identify new glass products and new markets.

Energy Sources

Our focus, relating to energy, has primarily centered around: how to use less of it! That's something we will probably never stop looking at.

A recent Kiplinger Letter says: "With high fuel prices here to stay, conserving energy is key to saving money for firms. Historically, industry has been wary of making heavy investments in energy conservation. Prices often fell as fast as they rose. But the energy outlook is much different now. Demand for energy will continue to outstrip supplies this year and next, and there's very little new production capacity being added."

So, seeking energy efficiency must be an ongoing effort.

Another approach, though, given the continual rise in the price of natural gas in recent months, and the likelihood that those increases are, for the most part, here to stay, is leading us to consider the possibility of alternative, cheaper energy sources. Much has been published in recent months about "Coal Gasification" and "Clean Coal". This plentiful fuel source has gone from being considered a serious polluter to one that can be processed economically to generate electricity releasing only carbon dioxide, which can potentially be captured and/or sequestered. For the last several months, the GMIC has been investigating the possibility of using this fuel directly in a glass melting furnace, bypassing the electricity generation stage to increase efficiency. Concerns in the past have related to the fact that coal-based "synfuels" have a lower BTU rating than natural gas. When burned in a high-oxygen atmosphere, however, slight modifications to the combustion process can make the system viable. We believe that such synfuel can be provided at prices of $4.00 to $4.50 per million BTU compared to much higher prices of $6.00 to $10.00 and over for natural gas, depending on location and time of year.

Systems range from 10 million BTU/hr up to 1,500 – 2,000. Lower costs are, of course, attainable with bigger systems, and the GMIC is in contact with the other industries involved in the IOF program to explore the possible establishment of a "network" of pipelines to bring synfuel from a central producer to industrial users in the same general area.

Refining Systems

As you surely are aware, two current research projects underway at this time involve the development of innovative melting systems: a submerged combustion melting system and another system using plasma melting. If successful, these systems will provide molten glass in a timely and cost-efficient manner. The melt, however, is likely to contain seeds and require further treatment to be able to be formed. The refining stage will be a separate function in this system. So, in parallel with the development of the melters, starting with a workshop held in May, a group of glass industry specialists are working to identify the appropriate refining technology to use with the new melting systems. At present, under consideration are 4 options: vacuum refining, helium gas, sonic and centrifugal. All have advantages and all have challenges. Members of our Production Efficiency Sub-committee, chaired by Stephen Gallaher of Owens Corning, are reviewing these factors with the goal of developing a proposal that will lead to research confirming the effectiveness of one system or another. We expect that, combining a new melting system with the appropriate refining system, we will see the introduction of what we've been referring to as the "Next Generation Melting Systems" that will open up new opportunities for producing glass at lower costs, with less energy, more flexibility and better control.

Developing New Relationships

The primary partner for glass industry development has been the Department of Energy since the early 1990s. We are extremely appreciative of the support that this government agency has given to our industry. The development of the oxy-fuel system for melting glass was largely supported by DOE cost-shared research, as was the development of a number of other systems that are in general use in the industry today. However, to be considered by the DOE, research carried out must be expected to lead to the reduction of energy in the melting process. As you all know, there are many other areas for possible improvement in our industry. So far, there have been no alternative partners to support work in the areas of basic knowledge of glass formulations, glass strength, forming processes, new uses for glass, composites optimizing the features of glass with other materials, and a number of other areas.

Department of Commerce

Since 2000 we have been developing relationships with the other industries that make up the DOE's IOF program. We've come to find that we all have similar concerns regarding our industries' status and prospects. We've been exchanging ideas regarding technologies and how best to work with the DOE. We have been discussing the possibility of developing programs with other federal agencies, and have agreed that cooperation is more likely if we approach them as a group of industries, rather than as a single sector. Late last year we made an initial visit to the Department of Commerce to propose the development of a program that might work in a similar way to the DOE's IOF Program. The DOC has established a new office for supporting U.S. Manufacturing and the "Industries of the Future" associations have formed ourselves into a new group to work with other organizations: the "Alliance for Materials Manufacturing Excellence" or AMMEX. In early September we presented a "White Paper" proposal to the DOC and are currently working with them to shape a program that could assist our industries to develop technologies in new areas. Here's a summary of some of our proposals:

- That the goals of the research be 1) improved productivity, 2) improved properties and performance of materials, 3) materials manufacturing processes that replenish or enhance the environment rather than deplete it.
- The justification for a project may be based in the benefits that occur anywhere in the value chain for that material and its products [e.g., the advances can occur in a materials manufacturing plant or in a plant that manufacturers goods from materials]
- The proposed research is widely applicable to a materials sector rather than to a company in a specific sector
- The non-federal partners must cost share at least 30% of the total cost of the proposed research
- Research proposals are evaluated and selected by DOC and materials manufacturing experts

We have also proposed cooperation to:

- Make permanent the Research and Experimentation Tax Credit
- Develop a comprehensive energy plan to encourage conservation and improve infrastructure
- Develop an "Interagency Working Group" that would bring together the needs and resources of several Federal Agencies
- Strengthen education, retraining and economic diversification
- Number of other issues.

We are working with several other organizations in pursuing this relationship with the DOC: the National Association of State Energy Officials (NASEO), the American Council for an Energy-Efficient Environment (ACEEE) and the North-East Mid-West Institute – a coalition of representatives and senators from 28 states seeking to improve job conditions in their states. We will keep our members informed as this relationship develops.

Environmental Protection Agency

Another initiative that has recently gotten underway is a plan within our Environmental Sub-Committee to interact with USEPA and individual states on implementation of the Routine Maintenance Repair and Replacement regulation that could impact the regulatory status of rebricking projects and other projects that improve energy efficiency. Committee members are discussing challenges they face in adhering to regulations, region by region, as well as with regards to Federal regulations. Our intent is to develop communications with the EPA through which we will be in a position to assist that agency to address environmental issues while encouraging glass companies to use best available technology without suffering serious adverse economic penalties. The sub-committee is just getting underway with an early round-robin of discussions. Pat Pride of PPG Fiberglass, and the chair of that committee, is leading this effort.

That's a general overview of some of the new directions the Glass Industry is taking under the umbrella of the Glass Manufacturing Industry Council. We have many challenges before us, but, with your continued support and involvement, we are convinced that the industry can move steadily towards a more prosperous, exciting and challenging period in our history.

Joining Organic Materials with Inorganic Glass—Future Composite?

John T. Brown, Glass Manufacturing

Introduction

In 1909 Edouard Benedictus, a set and costume designer for a French Theatre, patented Triple X™. A laminate made from two glass sheets bound to an inner layer of clear plastic. Today, every car, truck and bus built in the United States has a windshield of laminated safety glass.[1]

Another example that we can't prejudge the source of the breakthroughs for glass.

Today architectural glass stands beside the safety windshield as a fine example of combining polymers with glass to achieve safe applications that are also beautiful. Can you imagine a major city sky line today with out Glass covering the buildings exterior. Reflecting the clouds and sky in varying colors from clear to shades of blue to copper or even black.

The use of polymers to enhance the basic properties of glass has extended the use of glass. One attribute of glass that is barely touched is its intrinsic strength. As we will see, commercial glasses are exhibiting only a fraction of the potential strength, some say less than one two hundredth's of its possible strength.

For years people have been searching for ways to take advantage of glasses tremendous strength. Nearly 350 years ago, Prince Rupert of Bohemia brought curious glass drops to the Royal Society of England, in 1661. They had been made by dropping molten glass into cold water, the tadpole-shaped drops could be hammered repeatedly on their heads.

Why were they so strong?

Two hundred years later scientists discovered what was happening. When hot glass chilled, the fast cooling outside layer compressed. The slower-cooling inside is under tension. The only way to break the glass is to penetrate the tough outer layer and release the internal tension.

In the 1920's Corning physicist Jesse Littleton experimented to find the best temperatures for tempering glass and maximizing its strength. His discoveries gave glass makers control over the tempering process and laid the ground work for the flat-glass tempering method later developed by Saint-Gobain.

More recent innovation in glass strengthening has ranged from chemically exchanging smaller alkali ions for larger ions, stuffing the surface structure into a compressive outer layer. Examples range from frangible glass for circuit boards for secure electronics to medical serum delivery applications. Combining a high expansion interior glass with a low expansion glaze exterior will result in a very well protected compression layer which was used for Corning's discrete resistors and for their low cost dinner ware sold as Corelle® dinner ware. In many instances, strength by soaking in chemical baths or providing complex forming processes to combine glasses of different expansions prove too costly to the purchasing public. Our search for the solution to strength must be back to basics in understanding the cause for glass loosing the tremendous strength that is inherent in the covalent bond strength of oxygen and silica. By understanding this available strength we can delve into the mechanisms that strip this strength after the first moments of production.

Many pure glass manufactures have no use for polymers as they see polymers and aluminum, or metal as competitive materials replacing their basic business of providing containers. Perhaps by reviewing previous successes of polymers and glass we can begin to see paths that will grow glass, a valuable segment of the US economy—to allow glass to survive and thrive.

Before doing this, let's look at two obstacles causing concern for a secure future in The US Glass Industry.

Two Basic Hurdles to Overcome to Secure Our Glass Future

Capital Cost. to manufacture glass articles—and low return on investment—is highlighted as a major hurdle to continued growth of the Glass Industry.[2]

Availability of Energy. A new cost burden is energy. Society enjoyed low cost energy, to excess, until the 1973-74 oil embargo. Following this monopoly driven anomaly, real supply/demand pricing is indicating availability as perhaps a greater concern than cost. Global oil extraction will peak fairly soon. Estimates 3 range from 2006 to 2012. There remains as much oil in the ground as has been extracted—only it is more difficult to remove as a higher percentage of the last barrel is required to extract the next barrier. All energy sources follow oil on a cost per energy unit, to some extent, and more with the recent rise in oil cost.

Price will continue to increase but availability will loom as the basic threat to continuing as we are.

There is hope for the first two hurdles.

First, capital costs are being addressed by Next Generation Melting (NGM)—specifically: Submerged Combustion Melting—driven by a consortium of five glass manufacturers, a diverse base of funding, and managed by GTI. Fred Quan.

leading the Innovation Committee years ago, challenged us to make a 90% reduction in the equity invested to melt glass. This high risk investigation is supported by both public (DOE) and private funding.

Second, available and affordable energy for industry is looking for a bridge to the hydrogen economy, which, according to many analysts, is two to four decades in our future. Clean coal gasification is being demonstrated by our DOE in two installations. One in Tampa, Florida burning 2,400 tons of coal per day and another in Terre Haute, Indiana burning 2000 tons of Pet Coke per day. Both appear as a workable, affordable alternatives to natural gas, producing clean syngas in the neighborhood of two billion Btu's per hour, or burned in a GE turbine produces 185 MW. US coal reserves are estimated at 200-300 years at current energy usage rates—but this source for a clean gas for the Glass Industry is a long way from reality. Glass needs to forge alliances with all large industrial energy users to achieve results.

A New Golden Age for Glass

Most of us will agree that the US Glass Industry is in a perilous condition—-however, the core material, GLASS, is magical.

After five millennia, radical new uses for the ancient material, glass, keep coming to market.

Just ten years ago did you have a laptop computer,—wireless—with internet access in your home? Or a flat screen TV —with a choice of movies from a digital dial up phone. While the viewing face is glass, photonic communication, through glass fibers makes instant information retrieval possible. Glass fiber has replaced the long haul copper wire based tele-communications industry within the past ten years and given us bandwidth unimaginable only a decade ago. Memory in computers is the cheapest thing you can purchase due to chip density which is a result of shorter wave length glass lenses in stepper cameras. Life sciences, in support of affordable health are being stretched to do more with less, and glass is a key to continued success. These plus many others applications would not be possible with out special glasses using elements from the four corners of the periodic table and stretching forming science to new achievements.

Perhaps these would be better characterized as non-conventional glass, but let's take a peek into a possible future for conventional glasses.

Traditional glass, like containers, textile-fiber and wool-fiber, laboratory, table ware, architectural and automotive glass, to name a few, will enter a new GoldenAge if we unlock the secrets of brittleness. There is no reason why today's glasses could not be ten times to one hundred times more than the strength we accept today.

What do we need to spur us on to achieve this grand challenge? A prize of ten million dollars like the Ansari "S" prize and a materials scientist of the caliber of aviator Burt Rutan— to carry us across the strength threshold?

Observations on Strength

Nearly all glass products are cooled through their transition to room temperature in air containing water vapor. Sol-Gel may be an exception in that it avoids high temperature processing.

What these glasses have in common is increasingly higher tensile stress on the surface as they develop a thermal gradient of low temperature on the surface and higher temperature in the core. Below the glass transition temperature, internal stress relief due to fluid re-arraignment of molecules has essentially ended, in any meaningful time frame.

Further cooling introduces reversible stress which is finding ways to relieve on cooling by breaking bridging oxygen-silica bonds at the surface and over laying the surface with micro flaws. Internally, the glass has abundant alkalies and hydroxyls that can transfer and participate in this form of stress removal.

More importantly, water is available in ambient air. Typical dry winter weather contains 1000 parts per million (ppm) water, and humid summer air contains over 10,000 ppm. All seasons offer glass ample opportunity to relieve stress by incorporating hydroxyls from air to break bridging bonds.

I want to give credit to private communications on the subject of glass strength with Bill Prindle, Carlo Pantano, Charles Kurkjian, David Pye, Helmut Schaeffer, John Helfinstine, Suresh Gulati and Bulent Yoldas. Perhaps special thanks should be made to John Helfinstine, Suresh Gulati and Bulent Yoldas in helping me develop an understanding of how glass is damaged and the possible path to higher strength.

Below are typical breaking stresses for common glasses.

TABLE 1. BREAKING STRESSES OF ANNEALED GLASS, SHORT-TIME FLEXURE TESTS IN AIR[4]

Condition of Glass	lb/square inch
Surfaces ground and sandblasted	1,500-4,000
Pressed articles	3,000-8,000
Blown Ware:	
Hot iron molds	4,000-9,000
Paste molds	5,000-10,000
Inner surfaces	15,000-40,000
Drawn tubing or rod	6,000-15,000
Window glass	8,000-20,000
LCD (0.65mm)	~45,000*
Annealed fibers	
Annealed	10,000-40,000
Freshly drawn	30,000-400,000
Telecommunications fiber	> 100,000**

*Due to the thickness, LCD has never been measured as a bar, but as sheet glass using ring on ring. This data was not taken from McLelland and Shand but is Corning Inc. data, previously published.
**Corning Incorporated advertised strength for special fibers.

When reviewing this list of strengths of glass, notice that the inner protected surface of blown wear is much stronger than the outside. LCD glass is much stronger than window glass. LCD glass does not see any contact during manufacture. This begs the question of what is glass's potential strength.

Theoretical strength of the Si-O bond is calculated to be just over two million psi. Bureau of Standards testing shows that strength is not correlated with composition as much as with surface condition. Griffith flaws created in manufacturing contribute to enormous reductions in usable strength. In vacuum conditions, freshly drawn pristine fibers have been measured at 2,000,000 psi. In the above table, telecommunications fiber is outstanding in strength when compared to many glasses.

What is different about glass fibers for telecommunications?

- The core blank is stripped of water, well below detectable limits and the surrounding pure silica cladding is stripped of much of its water down to a few parts per billion.
- No alkalies
- Dry core is induction heated and fiber is drawn in a dry thermally conductive atmosphere.
- Rate of cooling is thousands of degrees Celsius per second.
- "Polymer" coatings are applied in a fraction of a second after the fiber is pulled from the blank to a final glass diameter of 125 microns, glass/polymer 250 microns.

There, the first mention of polymers! or Plastic—an important, maybe critical partner to the future of glass.

To go forward we need the material properties of polymers to enhance the properties of glass. Remember, glass is not alone in the energy availability and cost crunch—plastics are derived from hydrocarbons—100%. Sand, lime stone and soda ash are plentiful and local.

In a composite of the future, the backbone will be glass for strength, and polymer for protection—on the outside. A potential container of the future may be 50% less glass, with a thin coating of a few thousandths of an inch of polymers. This thin layer would provide all the color requirements and perhaps even advertising and allow for only clear flint glass. A benefit to recycling.

Attempts have been made in the past, and failed for a variety of reasons, primarily due to polymers missing requirements for recycling. This example of two bottles with matt surfaces represent a Wiegand Glas attempt to apply a 5 micron polymer coating on glass that is reduced in weight by 40%. At an "International Partners in Glass Research" meeting, I was given two one liter Coke bottles, not the ones pictured. We shared these bottles in a later meeting of the US Glass Industry, sponsored by GMIC, at DuPont's central Research Labs in Wilmington, DL, in March of this year. Later, we visited Wiegand Glas in a group that was traveling on to see the Submerged Combustion Melter in Belarus. Wiegand Glas is located in Steinbach, Germany, south of Berlin. Manfred Schramm, Director of R&D showed the data that confirmed no loss in strength with the 5 micron coating on a 40% weight reduced container. While the strength was equal to a full weighted bottle, the composite bottle felt like a PET bottle based on weight. In the March GMIC meeting, the two Coke bottles were given to DuPont.

Listening to remarks made by those that handled the one liter coke bottles we could grade the Wiegand effort as:

OK—Glass bottle

OK—Coating process

X—Polymer

Better surface characteristics are needed.

In general the plastic or polymer coating for the container industry must meet these additional needs.

- Easy separation of glass from polymer prior to crushing-
- Behave as a fuel upon entering the furnace-
- Produce only water vapor and carbon dioxide as a by product of exothermic combustion

Telecommunication fiber already includes polymers as an integral part of the product. A 250 micron fiber has 125 micron core. Because area is related to diameter squared, this by volume is a 75% polymer product. All the value as an information tool is the 25% core glass material. Really, nearly all the information is carried in only the center six microns of this glass fiber. However, there is no product without the protection provided by the polymer coating.

Life sciences are heavily invested in glass. This plastic optical grating on glass has the potential to automate laboratory measurements through optical interrogation. The ability to automate is enabled by physical properties of the polymer and glass.

The presentation included several slides provided by DuPont Central Research to show the connection between glass and plastics. I am indebted to Jeffrey Granato, "Global Architectural Marketing Director", for the visuals used in presenting this paper in Columbus, Ohio. Also, to Gary Turner of DuPont's Research Planning, for organizing our meetings and introducing, "we glass people", to many wonderful researchers at DuPont. In return, a tour of one of Saint Gobain's container furnaces was arranged by Tony Cappellino, Sr. VP, Technology, Saint-Gobain Containers. This offered a glimpse of a typical container plant, for the first time, to several DuPont researchers.

The following is an attempt to describe the slides shown in Columbus Ohio, used to emphasize the need to cooperate with the polymer industry.

A series of slides demonstrated the classification of heat strengthened, laminated or tempered glass and picture comparison of breakage to the variety of different strengthening processes. No industry has demonstrated a better adoption of polymers to float glass than the automotive industry in protecting the public with safety glass.

Combining new polymers, SentryGlas® Plus with side lights for automotive glass provides a material that is practically thief proof. To make the point concerning auto theft, a slide taken from a movie simulating a break in shows the thief using a baseball bat, then a crow bar, and finally gives up after making a small hole in the side light. If you want to imagine great things for glass what can be more imaginative than a glass bridge. In a composite picture showing the Dover, DL. race track, an actual glass bridge was constructed for viewing and walking over the race track.

There have been two NASCAR races since installation of the bridge. The test designed by the track committee was a cannon firing a hex lug nut at 170 mph into the insulated multiple glass membrane, that is the bridge.

Hurricane prone Florida can do away with plywood window protection if a DuPont inner polymer layer is sandwiched between two pieces of glass. The beautiful Broward Performing Arts Center in Ft Lauderdale, FL was completely re-glazed after construction with the hurricane protected glass. A demonstration with a pine two by four, shot at the front door at 35 miles per hour (50 fps) was used to demonstrate the resistance to force. The door remained intact but had to be replaced.

The new Reichstag Parliament building in Berlin, Germany has openness as its theme. The dome is glass and allows visitors to walk on glass path ways and peer down to observe parliament. The mirrored panels reflect or automatically focus light to control temperature.

Glass staircase and floors and walls in the Apple computer Flag Ship store in NY, City demonstrate beautiful new uses for glass.

Twin Towers in Kula Lumpur, one of the 5 tallest buildings in the world, represent the continued use of an old product in new architectural applications.

For fun, a short movie that had been made by British Telecom, showing importance of strength in glass in large aquarium's was shown. We are using clips to highlight a student contest on, "What could be expected of glass if strength were not a constraint".

Conclusion

Phil Ross sent me a beautiful history of furnace inspection meetings begun in early 70's by Jack Swearingen. It seems to fit the belief of many of us that to advance the material benefits of glass products, glass people must consider alternative materials to work with glass for improved products and performance.

Jack believed that glass manufacturers should compete by technology "leap frogging"—where each manufacturer made advancements by understanding improvements in the industry. They didn't just try to duplicate them, but to build on the results and find the next advancement. By this method, the industry would truly advance and more effectively compete with the real competition—Alternative Materials.

We need to think beyond our company and be willing to advance the industry. All the easy problems have been solved. But the difficult problems, when solved, will bring the greater rewards.

REFERENCES

1. Corning Museum of Glass; Technology Innovation section on flat glass.
2. Phil Ross and Gabe Tichner, *Technical and Economic Assessment for the US Glass Industry:* 2004
3. Richard Heinberg, The Party's Over (New Society Publishers,2003), p.118
4. George W. McLellan and E.B. Shand, *Glass Engineering Handbook,* Third Edition. p.6-3

Challenges to the US Glass Industry-Will the US Manufacture Glass in 2020?

Warren W. Wolf
Dr. W. W. Wolf Jr. Services, Ohio

Introduction

I shall begin with a positive statement: in 2020, and in the years beyond 2020, glass will continue to be in large scale world-wide use. Glass is too important a material to see a significant decline in its overall use. Glass is truly the most aesthetic of materials. It captures the light of our world and highlights a fine red wine in a glass, or allows windows that can open up our rooms and provide external views that bring that light to us. Similarly glass has the ability to be very hygienic and can be readily sterilized, accounting for its importance in laboratory and medical glassware. Glass can be produced in so many forms and shapes, from large glass plates that glisten in our skyscrapers, to fibrous forms to create fine insulation products which provides durability so they can be retained in walls and attics for years. In another fibrous form and when properly handled and coated, glass provides the most economical reinforcement to plastics and other matrices, so that low weight and high strength products are produced. Glass will continue in these and in many other forms to be useful and competitive against other materials.

In the next 15 years I can envision glass penetrating new uses in at least three areas. The telecommunication revolution is just beginning and is nowhere near its growth peak and in the next 15 years that revolution in telecommunications will produce a myriad of uses and conveniences to us. Somewhere in that 15 years and beyond we should also see an even larger shift of telecommunications from using electrons to photons as the main method of sending information to each other and that shift to photons or photonics should involve glass in ways not yet entirely foreseen. I also believe that the hygienic nature of glass and the ease of sterilizing its surfaces will also allow it even more uses as the bio-medical and biotechnology areas make further advances. A third area of increased innovation for glass applications is the enhancement of glass surfaces by coating and other processes so as to significantly provide new property capabilities.

We should never forget that the molten nature of glass allows us to form an infinite variety of material shapes. I know of no other material with such a unique ability to form different templates for final use. For these and other reasons I believe glass will be more widely used and manufactured than today in 2020. One mind set we must change is our tendency to label glass a commodity material, as a myriad of new uses certainly seem feasible. I state that only someone who lacks imagination can make the claim that glass is a commodity material.

However the question concerns the US and the manufacture of glass in the US in 2020. In the succeeding sections of this paper I will try to address the major concerns that can affect whether glass manufacture remains prominent in the US. To do that we need to consider four major issues of concern and then try to draw conclusions as a whole for the question.

- The transition to globalization, a service economy, and effects on US manufacture of glass,
- The challenges to the glass industry from rising energy costs and global warming,
- The need to maintain and increase innovation within the US, and
- Specific actions the US glass community should take.

Globalization, the Service Economy, and Future US Glass Manufacture

We can not predict the future but we can look at trends and make reasoned guesses. The first trend which is obvious is that employment in manufacturing will continue to decrease and there will be a greater shift of US employment to the service industry by 2020. This does not mean that US manufacturing of glass needs to decrease in terms of output. I quote from an article by Geoffrey Colvin in the May 3, 2004, issue of *Fortune* titled "Bush vs.Kerry: Who's stupider on jobs?" Ignoring the provocative title some interesting statistics are given. In 1990 the US had 169,000 steelworkers and 11 years later in 2001 the number had dropped to 88,000 or a 48% drop. But while 81,000 jobs disappeared the US produced 17% more steel. During this same time period US manufacturing sales grew to $4.3 trillion in 2001 from $2.8 trillion in 1990. During this time period as population grew, the total manufacturing population fell 7%.

If you wish an analogy, look at agriculture, and back in about 1850, 60% of our population worked in agriculture. Today that number is just over one percent. And yet the US does not lack in food production. And what drove the agriculture change is also what will drive the continued drop in manufacturing employment, productivity driven by technology innovations.

Now I am not saying these changes in employment statistics will not cause socio-economic issues, as they will, but those require considerations outside this paper. In particular it is probably no longer possible for a person to only have a high school education and go to work in a US manufacturing facility and assume by hard work he/she will have a lifetime where good wages are paid and they can raise a family and be able to afford a reasonably good life in America. For the truth is most service jobs, where employment is growing currently, are significantly lower paying, unless you have a college level education or higher

But if we wish to keep strong US glass facilities we must encourage them to promote more technology innovations and more productivity advances and most of these changes will decrease the labor component and not increase it. I see no other choice. For the US glass industry suffers from two problems as it competes world-wide. First, its labor rates are too high and to stay competitive it must utilize more controls, sensors and other productivity tools including artificial intelligence, to increase its productivity. This will not necessarily regain prominence for the US and may only keep us even with the rest of the world, for in the rest of the world, including China, we are also seeing less employment for more manufacturing output. The second item of importance is as the US struggles to reinvest in its glass facilities, their capital expenditure is too high for the pay back received. This is perhaps an even more major competitive problem and requires focused efforts by the US glass industry to change this picture by innovation around the area of capital expenditure. How to get more value output with less capital input can be approached by increasing either the value output or decreasing the capital input through innovation.

I will make only a few comments about globalization and its impact on US manufacturing. First there are probably no true US-only glassmakers. And glass making and its technology are a global phenomena and as such within the US we will face competition that will only increase and not decrease with time. The period after World War II when the US was dominant in glass making is gone and probably forever. To remain competitive, however, the US must continue to be a center for both innovation and even more importantly entrepreneurial leadership. In the next section we will talk some more about these two critical items. But in the next section I want to turn to two critical problems and see if they also present opportunities—rising energy cost and global warming. These are two important trends for the US glass industry to confront if it is to stay vital.

Two Challenges and Two Opportunities for the US Glass Industry/Rising Energy Costs and Global Warming

I should state at the beginning that I believe the world will see the effects of oil price increases very soon, perhaps in the next 3-6 years and certainly by 2020. I made a statement to this effect back at a USDOE/GMIC Energy workshop that preceded this Glass Problems Conference two years ago. This is of special importance to the US glass industry, because other than aluminum, glass manufacturing is the most energy cost intensive of all the primary high energy using businesses in the US. I also believe that global warming is real and that implications of meeting its effects could be the leading problem of this century for the world. Even if you do not believe my statement the political implications of global warming are now real and glass as an energy intensive industry must realize these political implications will have an influence on glass manufacture. Fortunately many of the same moves to decrease energy also help reduce carbon emissions.

The June 2004 issue of *National Geographic* carried an article "The End of Cheap Oil" by Tim Appenzeller that offered several scenarios. There is an optimistic prediction by David Greene of the Oak Ridge National Laboratory using US Geological Survey data that says the world peak in oil occurs as late as 2040. (It should be noted that the US Geological Survey is also famous for predicting in the 1950's that the oil peak in the US production would happen in 2000 or so and not in the 1970's when that US peak did occur.) Less optimistic were predictions by David Greene at Oak Ridge National Labs using data from UK based Colin Campbell which predicts that the world peak will happen by 2016 and outside the Middle East by 2006. I should tell you I lean toward the pessimistic data. Greene also notes that his prediction models have a built-in optimism since it factors no political or environmental constraints on production.

I will move onto global warming but not without a brief discussion of the work of the late M King Hubbert who was a visionary and was, as is often the case, both revered and reviled. Now the world has tended again to forget him. Hubbert was a geophysicist at Shell Oil. He had the temerity back in 1956 to state that the days of plentiful oil were limited. He predicted the decline of US oil peak in the 1970's in 1956. Hubbert's main insight was: it is not really how much oil is left but how quickly and cheaply it can be extracted, especially from a handful of large cheap-to-produce oil fields. Hubbert's peak concerns a theoretical milestone after which the output from a given field slows and becomes more costly to produce long before the last oil is removed. As reported in *The Wall Street Journal* of May 19, 2004, Houston energy banker Matt Simmons has a forthcoming book challenging claims about the reliability of Saudi Arabia oil output.

Poring over technical studies from Saudi petroleum engineers he believes "We could be on the verge of seeing a collapse of 30-40% of their production in the imminent future and imminent means something in the next three to five years, but it could even be tomorrow."

No one knows with certainty when we will see a significant increase in extraction prices for oil but we know it is coming and it could be very near.

I believe that rising energy costs are not all bad for the US glass industry if they prepare for them. Much progress was made in reducing energy use in the glass industry in the US since the 1970's and much of that new technology was put "on the shelf" when energy costs did not rise over the last twenty years. In terms of imports certain glass sectors are relatively immune to cost pressure from imports due to the transportation costs to import them into the US. One such sector is glass fiber insulation. But rising energy costs will also increase transportation costs so in that sense the net effect will be to assist the US glass industry, assuming that industry has prepared itself through innovation. We will discuss innovation in our next section but first some comments about global warming.

I am not going to give you a litany about science groups who have endorsed the concept that global warming is a serious problem nor am I going to discuss the Kyoto Treaty on Global Warming. I do personally believe from all the evidence that I read that global warming is happening but what I can not decipher from all the data and opinions is how much is a natural process of climate cycles and how much of it is human made due to our emissions from commerce. The answer to that may not mean much, as it is certain that the human contribution may be enough to cause some transitions that otherwise might not happen, and that human contributions will increase any natural driven effect, unlike past histories on earth.

But whether you believe global warming is happening, or not, it now has a political momentum for good or bad and the US glass industry as a large user of fossil-based energy will also be considered to have a large relative effect on global warming. Consider two recent events: The first of these was the release of the Pentagon's unclassified report in the February 9, 2004, issue of *Fortune*. The article was titled "Climate Collapse". The article was a summarization of an October 2003 Pentagon report commissioned three years ago by Secretary of Defense Rumsfield. Rumsfield picked Andrew Marshall, a well regarded DOD planner, and asked him to consider what an abrupt climate change might really be like. Marshall asked Peter Schwarz, who formerly headed planning at Royal Dutch/Shell Group to help write the report. What makes this report so interesting is that it focuses not at a 21st century-ending scenario but looks at possible near-term effects by 2020. Admittedly the scenarios are meant to be worst case but they are chilling and include: A few key coastal cities such as The Hague are uninhabitable by flooding. Melting of the Greenland ice sheets leads to a collapse of circulation systems in the North Atlantic Ocean which disrupts the temperate climate of Europe and northeast US that are made possible by the warm flows of the Gulf Stream. Europe's climate becomes more like Siberia. Australia and United States are affected but not as much as the rest of the world. This is only a forecast of a worst case but it has added fire to the controversy of global warming. As an example of what we can expect in the future I found two stories in *The Columbus Dispatch.* A head line on July 21, 2004, reads "8 States go after utilities in Ohio" and then on July 22 by "Lawsuit intends to slow global warming". In essence eight states (California, Connecticut, Iowa, New Jersey, New York, Rhode Island, Vermont and Wisconsin) and New York City have filed a lawsuit to force five electric power companies to reduce their carbon dioxide emissions which are linked to global warming. I think ignoring global warming is something the US glass industry cannot ignore. There are too many signs that we need to be in front, not behind, on global warming as well as energy reductions and increases in energy efficiencies.

Before we move to innovation there is one way for the US to most easily address its energy problems and the problem of global warming. It is a solution I think will fail to be passed for political reasons, but rather than debate nuclear technology, renewable energy, energy from coal, how to convert to hydrogen etc., the US could go a long way towards addressing its energy policy and avoid a lot of individual bureaucratic decisions, which may not even be right, and just adopt a carbon tax to be introduced gradually over the next decade. Yes, such a tax could be judged regressive since it hits lower incomes harder but this could be offset by reducing payroll levies which are the most regressive taxes of all. I doubt if I will persuade the US Glass Industry to support a carbon tax but I think it would be a best way to gradually allow manufacturing industries to move gradually in the right direction rather than waiting for a sudden "shoe to drop" in the midst of a crisis. The effects on high energy industries like glass could also be offset by tax credits.

Another potential idea is to accept taxes or mandatory reductions with a world-wide emissions trading program. As an example if there were a 20% mandatory reduction required of each company in a sector it could meet the target on its own by becoming more energy efficient or by switching from fossil fuels to alternatives. But it could also buy reductions on an open market from others who have cut emissions more than required and hence have excess credits to sell. The world credit trading idea seems a must with either a gradual tax and/or a gradual cap on CO_2 emissions.

Innovation in the United States of America

Although there are specific things we will discuss that the US glass industry can do to foster innovation, innovation will also need a favorable climate in the US overall in the next 15 years if the glass industry is to succeed here. I think there are two items needed to keep innovation within the US. The first is a strong science and engineering base in terms of human resources available to the US. And the second is to maintain a climate for entrepreneurial activity in the US.

I am concerned there may be a weakening on both these scores and I am most concerned about the drop of skilled talent going into advanced degree programs in the physical sciences and engineering who will be the resources for future developments. The National Science Board to the President concluded in *Science and Engineering Indicators 2004* that the US continues to be a world leader in science and technology, but this leadership faces an uncertain future as a result of ongoing economic and workforce changes. For many years the US benefited from minimal global competition in science and engineering but there are now attractive and competitive alternatives. Foreign students with temporary visas make up a third of science students, but there has been a significant drop in the number of high-skilled-related visas issued as a result of the current national security environment. The brightest students may also no longer be coming to study in the US because Canada, the European Union countries, China and India are making greater efforts to keep their brightest students at home. The number of US citizens receiving Ph.D.'s has been flat for ten years while the number of Asian Ph.D.'s has increased rapidly with something like a 50-fold increase in China. The full report is available at http:/www.nsf.gov/sbe/srs/seind04.

It should be noted that there has been an increase using 2002 statistics in the number of graduate students in science and engineering programs in US colleges and universities. Graduate students reached 455,355 in 2002. This was the first data by the National Science Foundation/NSF since September 11. It was driven by the number of US citizens enrolling in programs. There was a six percent drop in first-time foreign students. Unfortunately the increase in US students appears to be related to the economic downturn that started in 2000.Future number should be watched closely. At this point we should be striving to promote more of our best talent into graduate programs that will resource the future development programs of innovation.

But beyond getting good talent into our graduate programs we must also create a climate that fosters R&D and entrepreneur activity in this country. One good example: I am writing this paper in August, 2004 and as of this writing the US Congress has let the R&D tax credit elapse. There is a good chance of an extension but this has traditionally been done for one or only a few years. Why can't our Congress have the wisdom to make it permanent or at least extend it for some significant time period like 10 or 20 years? Remember business thrives on certainty and yet our tax laws get written for each two year election cycle. But in all reality a lot more could and should be done to foster innovation at the national level. First we need some meaningful reform of tort laws, not so business can escape doing harmful activities, but to prevent frivolous law suits. That factor continues to hamper legitimate new undertakings in the US because of a hidden litigation tax and the fear of lawsuits in new unknown areas.

The US cannot stop driving towards higher productivity. That means the introduction of new controls and other systems that will decrease manufacturing employment must be encouraged and not discouraged. Likewise new innovations must be encouraged that add product value or productivity even as jobs are reduced because of these innovations. The US must continue to review and regulate the US industry but it can not permit rules to become confusing and burdensome. And if we continue to dislodge workers as manufacturing follows the employment trends of agriculture we must find ways to retrain and allow workers displaced, enough economic security, or we will become in danger of losing support for our capitalistic system.

The above will require a tremendous amount of balancing of all interests in the US. There will need to be balance as well on the global level. We need to ensure that the US retains sufficient manufacturing strength so it remains secure and we must continue to allow the world to have open trade within agreed upon standards. The later is an even tougher balancing problem but it must be done.

Innovations within the US Glass Industry

So it is very easy all the US Glass Industry has to do is face up to rising energy costs and further environmental constraints and create new innovation to reduce capital constraints, increase productivity, reduce and improve energy efficiency and add value to the final products! Of course this is at a time of shrinking R&D budgets and declining university programs in glass science and engineering. We have only a little control over the national agenda to improve the manufacturing agenda and also to improve the climate for research and innovation. But the glass industry can unite with its colleagues in other similar industries to try and promote the values and policies that make the most sense. I think this is a superb way to go but we can't let this turn into something like agriculture subsidies that have so distorted the agricultural system. We want a fair and level playing field but we do not need welfare as it is not good for us or our industry.

To do anything the US glass industry has to unite and begin a process of speaking up for itself, as a unit, and not as "Lone Rangers". Ben Franklin's words are still poignant in our context over 200 years later. "We will either hang together or we will hang separately." The US glass industry must show leadership as a unit and it must address combined innovations whenever possible. The industry can unite behind pre-competitive research but that sort of work in the Universities is useless if all you do is give money and then walk away from engagement. And I do mean engagement that states what are our real problems, such as capital constraints and now energy costs, and what sort of work can help fix these that can be done in University settings. And if we can't get satisfaction in Universities we should form our own pre-competitive research behind common problems.

This has been done most successfully with the programs recently funded by the USDOE/ITP or US Department of Energy/Industrial Technology Programs. These are industry led programs that have been funded with matching dollars by ITP and industry to look at programs that save energy but also can reduce capital constraints by either providing flexibility or reducing initial capital. Some of these programs include new front end systems using oxy-fuel firing or new high intensity melters such as plasma melters or submerged combustion melters with innovations using oxy-fuel systems. In each case we have at least two, and sometimes six US glass companies involved in research that can make significant impacts on the US glass industry. Vendors another key source are also involved in sponsoring these programs.

But this model can also be used on other outside programs not requiring US DOE funding and its special requirements. For example what is the one problem which could be improved across all four glass sectors: bottle, plate, specialty and fiber? My answer is the inherent limit on glass products due to either their brittle or weak nature, or both, that is developed as a pristine glass surface encounters other surfaces or atmospheres such as water. But this suggests that we could start with pre-competitive research on how we can treat a high temperature glass surface, so that flaws and waste in the forming process are avoided or significantly reduced (productivity and energy savings). And a lot more value would exist in the final products, because of less fear by consumers of glass cutting them when broken or enhance the use of glass products in totally new applications. Eventually the pre-competitive research would need to spin away from just pre-competitive research and form a research consortium of interested companies who could leverage their pooled resources. This is an example of a daring challenge to add value that the US glass industry must address.

In summary the US glass industry should exist in 2020 as it provides a vital material that the world needs. But the US needs to worry less about other countries gaining and do more to promote its own interests and especially it needs to secure continued leadership in innovation of importance to itself. But it can't do this if each company listens to *The William Tell Overture* and tries to play Lone Ranger. Maybe the best future analogy if the US glass industry is to be here in 2020 is to behave like a great jazz combo of many good players who all can solo but who also can come together to ensure the melody is heard. If we "hang separately," then we know the answer to the initial question. The only thing stopping us right now is our old habits and, yes, even fears. We must learn to engage with each other!

I would hope the Glass Problems Conference would continue to revisit this question over the next 15 years. Incidentally the US glass industry has already done a good job of describing an innovative path forward. Please see the *Glass Industry Technology Roadmap* issued in July 2002 under the guidance and direction of the Glass Manufacturing Industry Council and The Office of Industrial Technologies/US Department of Energy. Contributions were made by most of the leading US glass companies and its key vendors. Prioritization is a requirement and some of that has started and can be continued by pooling resources around best next innovations. Again we must engage each other to set priorities and programs to make that happen!

9 781574 982381